# Advances in Experimental Medicine and Biology

Volume 1028

**Editorial Board**
IRUN R. COHEN, *The Weizmann Institute of Science, Rehovot, Israel*
ABEL LAJTHA, *N.S. Kline Institute for Psychiatric Research, Orangeburg, NY, USA*
JOHN D. LAMBRIS, *University of Pennsylvania, Philadelphia, PA, USA*
RODOLFO PAOLETTI, *University of Milan, Milan, Italy*

More information about this series at http://www.springer.com/series/5584

Bairong Shen
Editor

# Healthcare and Big Data Management

*Editor*
Bairong Shen
Center for Systems Biology
Soochow University
Suzhou, Jiangsu, China

ISSN 0065-2598    ISSN 2214-8019  (electronic)
Advances in Experimental Medicine and Biology
ISBN 978-981-10-6040-3    ISBN 978-981-10-6041-0  (eBook)
DOI 10.1007/978-981-10-6041-0

Library of Congress Control Number: 2017950496

© Springer Nature Singapore Pte Ltd. 2017
This work is subject to copyright. All rights are reserved by the Publisher, whether the whole or part of the material is concerned, specifically the rights of translation, reprinting, reuse of illustrations, recitation, broadcasting, reproduction on microfilms or in any other physical way, and transmission or information storage and retrieval, electronic adaptation, computer software, or by similar or dissimilar methodology now known or hereafter developed.
The use of general descriptive names, registered names, trademarks, service marks, etc. in this publication does not imply, even in the absence of a specific statement, that such names are exempt from the relevant protective laws and regulations and therefore free for general use.
The publisher, the authors and the editors are safe to assume that the advice and information in this book are believed to be true and accurate at the date of publication. Neither the publisher nor the authors or the editors give a warranty, express or implied, with respect to the material contained herein or for any errors or omissions that may have been made. The publisher remains neutral with regard to jurisdictional claims in published maps and institutional affiliations.

Printed on acid-free paper

This Springer imprint is published by Springer Nature
The registered company is Springer Nature Singapore Pte Ltd.
The registered company address is: 152 Beach Road, #21-01/04 Gateway East, Singapore 189721, Singapore

# Contents

1 **How to Become a Smart Patient in the Era of Precision Medicine?** ................................................. 1
Yalan Chen, Lan Yang, Hai Hu, Jiajia Chen, and Bairong Shen

2 **Physiological Informatics: Collection and Analyses of Data from Wearable Sensors and Smartphone for Healthcare** ............ 17
Jinwei Bai, Li Shen, Huimin Sun, and Bairong Shen

3 **Entropy for the Complexity of Physiological Signal Dynamics** ... 39
Xiaohua Douglas Zhang

4 **Data Platform for the Research and Prevention of Alzheimer's Disease** ........................................................... 55
Ning An, Liuqi Jin, Jiaoyun Yang, Yue Yin, Siyuan Jiang, Bo Jing, and Rhoda Au

5 **Data Analysis for Gut Microbiota and Health** ................. 79
Xingpeng Jiang and Xiaohua Hu

6 **Ontology-Based Vaccine Adverse Event Representation and Analysis** ................................................... 89
Jiangan Xie and Yongqun He

7 **LEMRG: Decision Rule Generation Algorithm for Mining MicroRNA Expression Data** ............................... 105
Łukasz Piątek and Jerzy W. Grzymała-Busse

8 **Privacy Challenges of Genomic Big Data** .................... 139
Hong Shen and Jian Ma

9 **Systems Health: A Transition from Disease Management Toward Health Promotion** ......................................... 149
Li Shen, Benchen Ye, Huimin Sun, Yuxin Lin, Herman van Wietmarschen, and Bairong Shen

# Chapter 1
# How to Become a Smart Patient in the Era of Precision Medicine?

Yalan Chen, Lan Yang, Hai Hu, Jiajia Chen, and Bairong Shen

**Abstract** The objective of this paper is to define the definition of smart patients, summarize the existing foundation, and explore the approaches and system participation model of how to become a smart patient. Here a thorough review of the literature was conducted to make theory derivation processes of the smart patient; "data, information, knowledge, and wisdom (DIKW) framework" was performed to construct the model of how smart patients participate in the medical process. The smart patient can take an active role and fully participate in their own health management; DIKW system model provides a theoretical framework and practical model of smart patients; patient education is the key to the realization of smart patients. The conclusion is that the smart patient is attainable and he or she is not merely a patient but more importantly a captain and global manager of one's own health management, a partner of medical practitioner, and also a supervisor of medical behavior. Smart patients can actively participate in their healthcare and assume higher levels of responsibility for their own health and wellness which can facilitate the development of precision medicine and its widespread practice.

**Keywords** Smart patients • Precision medicine • Healthcare

---

Y. Chen
Center for Systems Biology, Soochow University, Suzhou 215006, China

Department of Medical Informatics, School of Medicine, Nantong University, Nantong 226001, China

L. Yang • H. Hu
Center for Systems Biology, Soochow University, Suzhou 215006, China

J. Chen
School of Chemistry, Biology and Material Engineering, Suzhou University of Science and Technology, No1. Kerui road, Suzhou, Jiangsu 215011, China

B. Shen (✉)
Center for Systems Biology, Soochow University, No.1 Shizi Street, Suzhou, Jiangsu 215006, China
e-mail: bairong.shen@suda.edu.cn

## 1.1 Introduction

Healthcare is undergoing a profound revolution as the consequence of precision medicine, which utilizes modern genetic technology, molecular imaging technology, and biological information technology, combined with patient's living environment, lifestyle, and clinical data, to achieve precision disease classification and diagnosis and develop a personalized prevention and treatment [1]. Meanwhile, with the development of smart medicine [2], more hospitals start to utilize various kinds of high-tech means, such as artificial intelligence (AI) [3, 4], gene therapy [5], sensing technology [6], etc., to achieve better and more ideal treatment level and outcome. In addition to the change of medical service and payment mode, as the center of current medical model, patients are faced with more requirements and pressure on how to face the complex multidimensional disease data [7], including clinical chemistries, molecular and cellular data, organ, phenotypic imaging, social networks, etc.

These changes and transformations are all presenting more challenges to patients in the new era, when precision medicine is emerging as a natural extension that integrates research disciplines and clinical practice to build a knowledge base that can better guide individualized patient care [8]. How to become a smart patient to adapt to the current medical model and to achieve precision health and wellness is pressing especially for patients with chronic diseases.

To date, the most relevant research about the smart patient is the book wrote by Roizen, M. F. and Oz, M. C. in 2006 – *You the Smart Patient: An Insider's Handbook for Getting the Best Treatment* [9], a how to guide for navigating common healthcare situations. Although quite a few of the later studies have also referred to the "smart patient," the definition and implementation are mixed and not quite absolute and thorough [10–15]. In recent years, e-patient [16–19] may be the closest study to smart patients. However, the definition and implementation of both "smart patient" and e-patient are still not clear.

Here, we put forward clear definition and meaning of a smart patient, summarized the existing foundation, explored the concrete realization model, and discussed the need for the conditions as well as the necessity and possibility.

## 1.2 Research Methods

### 1.2.1 Theory Derivation Processes: A Thorough Literature Review

We applied the idea of evidence-based medicine, and the method of systematic review. "systematic diagnosis," "participatory," "health application," "smart medicine," and "precision medicine" were comprehensively searched in PubMed with different combinations and analyzed according to different research purposes.

As it is a relatively new concept, there are little researches directly about the smart patient; we collected the relevant researches as much as possible for subsequent systematic classification and summary analysis.

### 1.2.2 DIKW: Data, Information, Knowledge, and Wisdom Framework

The classic "data, information, knowledge, and wisdom (DIKW) framework" in information science was performed to construct the definition structure and participatory model of smart patients. The DIKW framework is a hierarchy progressing from data to information, knowledge, and wisdom which has been maturely used in a variety of research areas [17, 20, 21].

We utilize this framework model to demonstrate how healthcare data is ultimately used by smart patients and how smart patients participate in medical care. The progression in each step of the framework is based on the addition of context to allow interpretation. Namely, data in context becomes information, information in context becomes knowledge [22]. Equally, the meaning of the smart patient is successively refined through the application of context.

## 1.3 Results

### 1.3.1 What Is a Smart Patient in the Current Medical Era?

Combined with previous researches and summary of the latest literatures, the definition of the patient is gradually coming into focus. To sum up, a smart patient is someone who can take an active role in his or her own health management: with the provided reliable health information to make evidence-informed choice, utilize diversified smart technologies to perform self-monitoring, self-care, and equal involvement in clinical decision-making, to get best and most appropriate treatment.

The ultimate goal is to inspire more patient participation and to achieve precision personalization and precision prevention and prediction. As the donor and recipient of medical development, the smart patient is the core driving force of the development of 4P medicine (predictive, preventive, personalized, and participatory medicine), especially about the specific implementation of precision medicine.

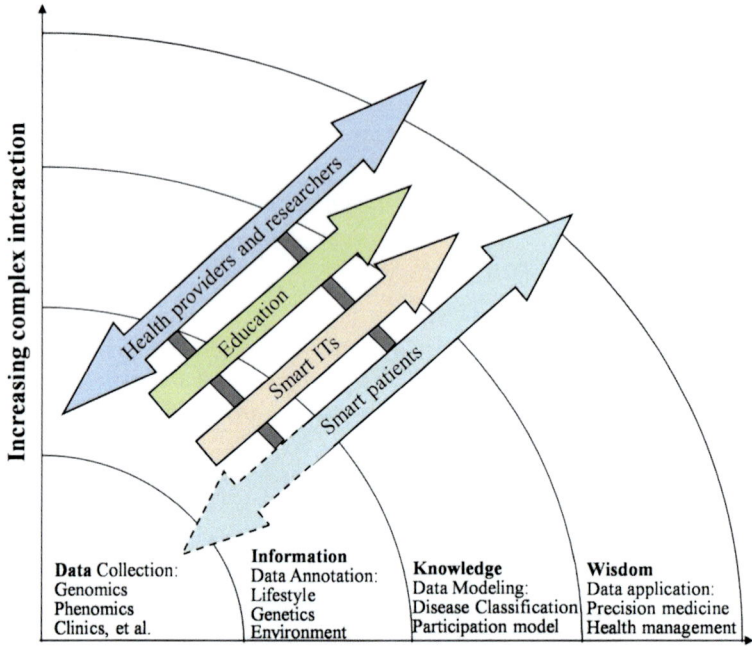

**Fig. 1.1** The DIKW definition framework and implementation model of smart patients

## 1.3.2 How to Become a Smart Patient?

### 1.3.2.1 The Theoretical Framework: DIKW Framework

We conclude that the DIKW framework furnishes a foundation for linking theory and practice of a smart patient. Figure 1.1 shows the definition framework and implementation model of smart patients established by DIKW method, which demonstrates the collaboration including smart patients, health providers, and researches as well as the function of smart ITs (information technologies). The interactions and interrelationships increase from left to right along the horizontal axis which also reveals the transformation of data to application. Complexity increases along the vertical axis. A smart patient, clinician, or researcher can move back and forth across the domains of DIKW. Each can traverse the domains alone with the auxiliary of smart ITs and education of the latest medical knowledge which also can facilitate the collaboration between them, potentially enhancing the development of wisdom in each. Each element of the framework is described in each layer.

### 1.3.2.2 Data Evolution and Application: From Patient Perspective

Over the last decade, many significant technological breakthroughs have revolutionized human complex disease researches in the form of genome-wide association studies (GWASs) [23, 24]. Investigators have begun to exploit extensive electronic medical record systems to conduct a genotype-to-phenotype approach when studying human disease – specifically, the phenome-wide association study (PheWAS) [25–27], a relatively new genomic approach to link clinical conditions with published variants [28]. This "translation" involves correlating genotype with phenotype, which often requires to deal with information at all structural levels ranging from molecules and cells to tissues and organs and individuals to populations [7]. Genotyping and large-scale molecular phenotyping are already available for large patient cohorts and may soon become available for many patients.

Research shows that about 5–10% of all cancers are caused by genetic defects, while the rest of 90–95% are caused by environmental factor and lifestyle, including unhealthy diet (30–35%), tobacco use (25–30%), and alcohol use (4–6%) [29]. The environmental agents that can interfere with DNA methylation are widespread and also depend on lifestyles. Smoking, alcohol consumption, UV light and chemical exposure, or factors linked to oxidative stress are some of the most common and important lifestyle aspects that may alter the DNA methylation profile [30, 31]. However, the problem is how to label and annotate these data.

As displayed in Fig. 1.2, patients' health-related data present a scene of diversification and complexity. If these data can be effectively integrated and utilized, patients' disease therapy and health management can achieve unprecedented accuracy. In the graph below, the problems need to be solved were set out.

In terms of data storage, patient level databases, in contrast to knowledge bases, are needed and not limited to aggregated data and contain individual patient genomic data; in addition, in some cases, they also contain limited or lifetime de-identified clinical data [32]. The focus of these databases is not a simple list of data, but to integrate different purposes to meet the different requirements of patients.

Classification and hierarchy of the data should be combined with different disease types. When it comes to disease classification, although there are many traditional and new disease classification methods, e.g., the International Classification of Diseases for Oncology-10 (ICD-10) [33], Health Level Seven International (HL7), and Systematized Nomenclature of Medicine-Clinical Terms (SNOMED-CT) [34], more accuracy classification of diseases are required in the precision medicine era especially from the patient perspective.

Similarly, the data analysis model, the application model, and visualization are all the premise of the realization of smart patients. We eagerly expect that along with the mature, consummate, and safety of these technologies, a smart patient can have more access to his own health state and acquire tailor-made individualized treatment plan according to the personal characteristics.

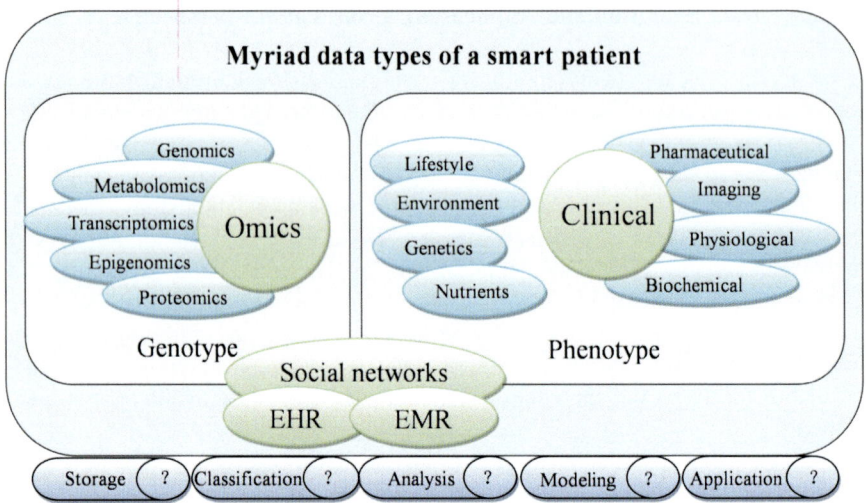

**Fig. 1.2** Data distribution of a smart patient to achieve precision health and wellness. Abbreviations: *EHR* electronic health record, *EMR* electronic medical record

### 1.3.2.3 Participatory Approaches of Smart Patients in Medical Procedure

As depicted in the schematic diagram of Fig. 1.1, with the rapid development of medical technology and systematic education, patients are able to monitor their personal health and have more in-depth understanding of disease more than ever before. Figure 1.3 lists four major approaches for smart patients to participating in disease and wellness management and medical research in the current medical era.

Self-Assistant Smart Diagnose, Treatment, and Disease Management

Smart IT is undoubtedly a critical facilitator to the creation of the smart patients. With wearable devices, patients, hospitals, Internet, robotics, and doctors will be in tandem with each other named new medical normalcy. Using the Internet, robots, sensing device and health Apps, etc., medical workers can also provide better healthcare for patients. With the aid of the present detection techniques, patients can determine their risk factors and whether they are genetic or due to lifestyle choices. A typical example is engineering of smart multifunctional theranostics which appears to be the next step for simultaneous diagnosis and therapy of cancer [35].

In recent years, health apps are more accessible for patients and have potential for both primary care practitioners and patients. Not only does apps help provide assistant treatment, diagnosis, and disease management but also can make patients

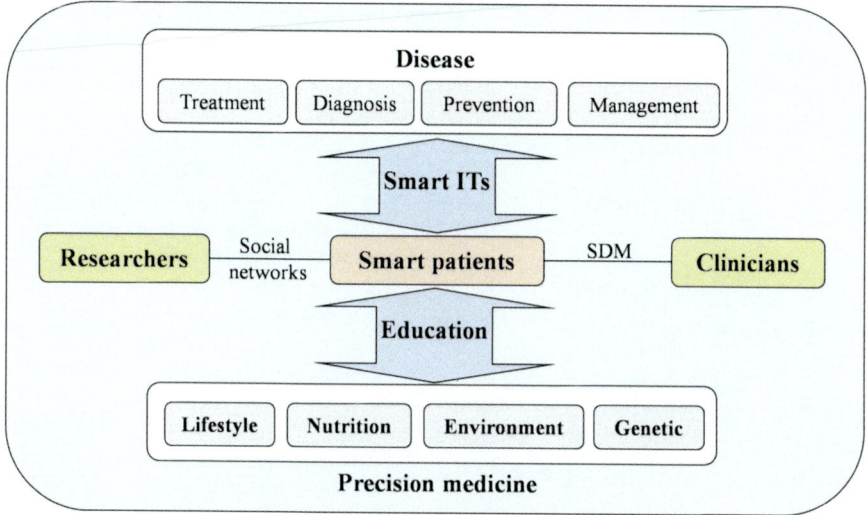

**Fig. 1.3** Participatory model of a smart patient. Abbreviation: *SDM* shared decision-making

better understand their own health status, strengthen communication with doctors, and participate in medical decision-making to achieve better medical effect.

In addition, long-term health management is challenging for the rapidly growing number of patients with chronic diseases. Interventions and auxiliary through smart IT may offer promising solutions and a completely or at least partially effective tool to assist in managing some chronic diseases [36].

The impact of mobile handheld technology on the work and practices of hospital physicians and on patient care has been summarized in recent reviews [37–39]. Here, we made a summary of the current researches about apps mainly for patients according to different functions, which are all supported by public publications.

"Applications, apps, and patient" were searched in Pubmed within the latest 5 years. And more than 10,000 studies were retrieved. However, partial apps were presented in Table 1.1 according to four function module including treatment, diagnosis, health management, and intervention functions.

While the health apps are making measure up, smart patients should be able to make smart choice and rational utilization of all kinds of intelligent means to encourage behaviors that protect health and assist self-health management.

Shared Decision-Making with Health Providers

Communication between the patient and physician is central to medical care. Effective communication can not only improve patients' knowledge about their illness but also can enlist them to be partners in their care, improve adherence to treatment, and improve satisfaction with care [70]. Educational interventions to

Table 1.1 Sorting and classification of partial medical applications (APPs) for patients

| Apps type | Apps name | Country | Introduction and function |
|---|---|---|---|
| Treatment | iTinnitus [40] | Australia | A sound therapy package for patients with tinnitus |
| | smartCAT [41] | United States | An m-health platform for cognitive behavioral therapy (CBT) of child anxiety |
| | ART app [42] | United States | An alcohol resilience treatment (ART) app in android system and an accessory of bilateral tactile stimulation |
| | Sleep Time [43] | New Jersey | Monitor sleep |
| | Medication plan [44, 45] | United States | Support the regular and correct intake of medication |
| Diagnosis | iTongue [46] | USA | ZHENG classification in TCM |
| | NeuroScreen [47] | United States | Detect HIV-related NCI that includes an easy-to-use graphical user interface with ten highly automated neuropsychological tests |
| | EncephalApp_Stroop [48] | USA | A short, valid, and reliable tool for screening of minimal hepatic encephalopathy(MHE) |
| | iFall [49] | USA | An application for fall detection and response |
| | Stroke vision [50] | United States | Assess for visual acuity, visual field, and visuospatial neglect, as well as novel tools for the education of patients, carers, and staff |
| Disease management | My cancer diary [51] | Korea | Cancer management |
| | ePAL [52] | United States | Empower patients for cancer pain management by prompting regular pain assessments and coaching for self-management |
| | Smart [53] | USA | A useable and feasible method for monitoring daily pain symptoms among adolescents and adults with sickle cell disease-related pain |
| | Diabeo [54] | France | A telemedicine solution for diabetes management |
| | Cardiomobile [55] | Australia | A real-time remote monitoring system for cardiac rehabilitation |
| | Pulmonary rehabilitation [56] | UK | An application based on standard pulmonary rehabilitation program for self-management, consists of Bluetooth pulse oximeter and smartphone |
| | Asthma peak flow monitoring [57] | UK | An application to monitor peak flow of asthma patients |
| | uHear [40] | Australia | A hearing loss self-assessment test |
| | Sleep aid [40] | Australia | A sleep apnea management application |
| | HealthPROMISE [58] | | |

(continued)

**Table 1.1** (continued)

| Apps type | Apps name | Country | Introduction and function |
|---|---|---|---|
| | | United States | Patient-engaged care that is centered on enhanced self-management and improved doctor-patient communication for inflammatory bowel disease(IBD) |
| | PD Dr. [59] | United States | Collects quantitative and objective information about Parkinson's disease(PD) and enables home-based assessment and monitoring of major PD symptoms |
| | SMART platform [60] | United States | Monitoring medication adherence |
| | Strong heart [61] | Korea | An educational learning instrument for coronary artery disease (CAD) patients |
| | PCHRs [62] | USA | Provide patients with a secure repository of their health information that can be exposed to apps across a programming interface |
| | "Interactive diet and activity tracker" (iDAT) [63] | Singapore | A caloric-monitoring app among patients with type 2 diabetes and managed in primary care |
| | My chart in my hand [51] | Korea | Personal health record |
| Health intervention | Purdue Momentary Assessment Tool [64, 65] | France | A human behavior monitoring tool |
| | Lose It! [66] | USA | A lifestyle intervention |
| | My Meal Mate [67] (MMM) | UK | For weight loss to be used on an android operating system |
| | SmartMove [68] | Ireland | Increase physical activity in primary care |
| | OpenLabyrinth 3.3 [69] | Greece | Medical education arsenal with capacities of creating simulation/game-based learning episodes, massive open online courses, curricular transformations, and a future robust infrastructure |
| | eCAALYX [37] | UK | A remote monitoring system for older people with multiple chronic conditions |

increase patients' participation in their visits with physicians have also been efficacious which reduce the information inequality between doctors and patients and finally achieve shared decision-making (SDM) [71].

One effective empowerment strategy in minority populations is storytelling, or narrative. Narrative shows promise as a potential method to empower minority patients with chronic diseases, by promoting self-care and facilitating SDM [72].

Nevertheless, interventions will play an important part in increasing smart patient participation in healthcare. The role of health professionals in supporting disadvantaged patients and tailoring information to their needs is essential.

## Participate in Medical Researches Through Social Networks

In the new era, an increasing number of patients are going online to access information about their health and talk to other patients with a shared condition. Many patients share advice and details about their treatments and symptoms with one another as well as researchers. Clinical trial researchers increasingly use the Internet for recruiting subjects, communicating with participants, and even collecting data [73].

A smart patient is in a position to clearly understand the value of their data and how it can advance biomedical science without injury to them and will be empowered to participate actively in patient (consumer)-activated social networks. Several such networks are already changing medicine, and one example is the "quantified self" networks that have now spread widely in the United States [74]. Individuals in these networks use digital devices to measure their own physical parameters (weight, pulse, respiration, quality of sleep, stress, and so on).

Main patient groups like the Life Raft Group for patients with gastrointestinal stromal tumor have successfully mobilized their members to study the effectiveness of investigational treatments [75], the accumulation of information on the patient's participation in the exchange, and the patient's perspective.

Besides, PatientsLikeMe is a Web-based community and research platform where patient members share details about their treatments, symptoms, and conditions, with the intention of improving their outcomes [76].

Moreover, the Cochrane library, 23andme, as well as WHO are all similar platforms for patients communication and safety maintenance. To achieve a participatory healthcare system, major technical and societal challenges have to be overcome, and this will require close integration with systems medicine and big data [77].

## From Data to Action: Patient Education Is the Key and Decisive Factor

In the current medical model, data is no longer a one-way input and output; patients play an increasingly important role in the whole process of medical development. As shown in Fig. 1.3, to allow the participation model to work effectively, we must first of all implement relevant education and guidance to smart patients, such as the development of medical technology and the application of the latest treatment means and ways of participation. On the one hand, this can enhance the patient's participation. On the other hand, the higher the patient's understanding of health, the more requirements are produced which in turn promote the development of medicine. In the current medical ecological environment, a smart patient should master self-control ability and prevention awareness, pay more attention to different environmental exposures, and learn to develop health lifestyles, regimens, and regular exercise to achieve better disease prevention or disease prognosis.

However, it is worth noting that who will undertake the education and how it operates are still the problems to be solved at present.

## 1.4 Discussion and Conclusion

### 1.4.1 The Importance of Smart Patients in the Healthcare System

In a health competent society, individuals, communities, and institutions should have the knowledge, attitudes, skills, and resources needed to improve and maintain health, and folks should act appropriately and consistently to the health challenges they face. Research shows that when patients fully disclose their concerns, expectations, and preferences, providers can assess their problems more accurately and offer better advice [78].

Integrating all the above results, obvious and decisive roles of the smart patient can be found in the progress of healthcare system. First and foremost, as the basis for the entire healthcare system, a smart patient can perform a better self-monitoring and health management which greatly improve the quality of primary healthcare. It not only saves healthcare resources [79] but also promotes the optimization of the healthcare system.

Patient participation in care delivery can have great benefits in increasing patient satisfaction, enhancing communication between patients and doctors which ease the current tense doctor-patient relationship. It also helps supervise the medical care and reduces adverse events [80]. As a member of healthcare system stakeholders, a strong sense of the smart patients can better inspire and motivate the development of healthcare.

### 1.4.2 Opportunities and Challenges

In recent years the general public has become more health conscious, due in part to network and wearable sensor technologies that enable the nonexpert to easily capture and share significant health-related information on a daily basis as part of attempts [81]. An increased prevalence of chronic illnesses and the identification of the benefits of team-based healthcare delivery are resulting in more care delivery by interdisciplinary healthcare teams (IHTs) [82]. A growing body of medical evidence come from the perspective of patients (or consumer) and the patient-reported outcomes [83], like off-label prescribing studies [84]. Copious social networks provide convenience for health education and management and also provide convenient ways to collect patient experience and resources. The popularity of various formats of health applications brings the patient health information on fingertips and supervision at any time. More importantly, the rapid development of the 4P medicine, especially for participatory medicine, needs more patients' participation and cooperation [85]. With the development of the above areas, the realization of the wisdom of smart patients is of unlimited potential.

Certainly, there are still a lot of obstacles and deficiencies on the way of the realization of the smart patient. Firstly, there are obvious questions relating to the ethical, legal, societal, security, privacy, and policy regulatory aspects of medical resources of medical institutions and individual patients, which have been discussed extensively elsewhere [86]. Secondly, there are also a lot of research questions about the application of intelligent methods in the medical field, such as the accuracy and precision of the apps, patients over rely on intelligent means which delay the medical treatment, etc. These measures are only meant to assist smart patients in health management and better participate in medical decision-making, rather than completely replace the traditional medical treatment; thirdly, the lack of disease and health education for patients still does not cause enough attentions and need to be addressed; the last but not least, due to the complexity of medical data, we need to systematically think about disease and well-being. However, there is lack of unified standard and analytical methods, especially the relevant system models, which are important problems to be solved.

### 1.4.3 Conclusion

To a certain extent, the doctor knows the average value and the general situation of disease, but patients know more about themselves and have more in-depth feel of their own illness and disease. In general, a smart patient is not merely a patient but more importantly a captain and global manager of one's own health management, a partner to medical practitioner, and also a supervisor of medical behavior against the common enemy – disease.

### 1.4.4 Practice Implications

As the donor and recipient of medical development, the smart patient is the core driving force of the development of precision medicine; they received systematic and comprehensive health education and can better adapt and promote the rapid development of medicine.

**Acknowledgments** This study was supported by the National Natural Science Foundation of China (NSFC) (grant nos. 31670851, 31470821, and 91530320) and National Key R&D programs of China (2016YFC1306605).

## References

1. Hood L, Flores M (2012) A personal view on systems medicine and the emergence of proactive P4 medicine: predictive, preventive, personalized and participatory. New Biotechnol 29(6):613–624
2. Soller BR et al (2002) Smart medical systems with application to nutrition and fitness in space. Nutrition 18(10):930–936
3. Giovanni Acampora DJC, Rashidi P, Vasilakos AV (2013) A survey on ambient intelligence in health care. Proc IEEE Inst Electr Electron Eng 101(12):2470–2494
4. Kartakis S et al (2012) Enhancing health care delivery through ambient intelligence applications. Sensors (Basel) 12(9):11435–11450
5. van der Werf CS et al (2015) Congenital short bowel syndrome: from clinical and genetic diagnosis to the molecular mechanisms involved in intestinal elongation. Biochim Biophys Acta 1852(11):2352–2361
6. Ona T, Shibata J (2010) Advanced dynamic monitoring of cellular status using label-free and non-invasive cell-based sensing technology for the prediction of anticancer drug efficacy. Anal Bioanal Chem 398(6):2505–2533
7. Chen J et al (2013) Translational biomedical informatics in the cloud: present and future. Biomed Res Int 2013:658925
8. Bahcall O (2015) Precision medicine. Nature 526(7573):335
9. Roizen MF, Oz MC (2006) You the smart patient: an insider's handbook for getting the best treatment. Free Press, New York
10. Zengota EG (1986) Planning a "smart" patient security system. Contemp Longterm Care 9(8):30. 32
11. Seidman S (1990) Press release: European community to use smart patient cards. J Med Syst 14(3):158–159
12. Park CS et al (2011) Development and evaluation of "hospice smart patient" service program. J Korean Acad Nurs 41(1):9–17
13. Kim YM, Bazant E, Storey JD (2006) Smart patient, smart community: improving client participation in family planning consultations through a community education and mass-media program in Indonesia. Int Q Community Health Educ 26(3):247–270
14. Hoo WE (2006) On "smart" patients as consumers. J Healthc Qual 28(6):4. 12
15. Hogan NM, Kerin MJ (2012) Smart phone apps: smart patients, steer clear. Patient Educ Couns 89(2):360–361
16. Abdaoui A et al (2015) E-patient reputation in health forums. Stud Health Technol Inform 216:137–141
17. Gee PM et al (2012) Exploration of the e-patient phenomenon in nursing informatics. Nurs Outlook 60(4):e9–16
18. Gee PM et al (2015) E-patients perceptions of using personal health records for self-management support of chronic illness. Comput Inform Nurs 33(6):229–237
19. Meehan TP (2014) Transforming patient to partner: the e-patient movement is a call to action. Conn Med 78(3):175–176
20. Cook DA et al (2015) A comprehensive information technology system to support physician learning at the point of care. Acad Med 90(1):33–39
21. Smith PF, Ross DA (2012) Information, knowledge, and wisdom in public health surveillance. J Public Health Manag Pract 18(3):193–195
22. Herr TM et al (2015) A conceptual model for translating omic data into clinical action. J Pathol Inform 6:46
23. Dorajoo R, Liu J, Boehm BO (2015) Genetics of type 2 diabetes and clinical utility. Genes (Basel) 6(2):372–384
24. Hebbring SJ (2014) The challenges, advantages and future of phenome-wide association studies. Immunology 141(2):157–165

25. Pendergrass SA et al (2011) The use of phenome-wide association studies (PheWAS) for exploration of novel genotype-phenotype relationships and pleiotropy discovery. Genet Epidemiol 35(5):410–422
26. Pendergrass SA et al (2013) Phenome-wide association study (PheWAS) for detection of pleiotropy within the population architecture using genomics and epidemiology (PAGE) network. PLoS Genet 9(1):e1003087
27. Denny JC et al (2013) Systematic comparison of phenome-wide association study of electronic medical record data and genome-wide association study data. Nat Biotechnol 31(12):1102–1110
28. Denny JC et al (2010) PheWAS: demonstrating the feasibility of a phenome-wide scan to discover gene-disease associations. Bioinformatics 26(9):1205–1210
29. Anand P et al (2008) Cancer is a preventable disease that requires major lifestyle changes. Pharm Res 25(9):2097–2116
30. Gorelik GJ, Yarlagadda S, Richardson BC (2012) PKCδ oxidation contributes to ERK inactivation in lupus t CELLS1. Arthritis Rheum 64(9):2964–2974
31. Romani M, Pistillo MP, Banelli B (2015) Environmental epigenetics: crossroad between public health, lifestyle, and cancer prevention. Biomed Res Int 2015:587983
32. Huser V, Sincan M, Cimino JJ (2014) Developing genomic knowledge bases and databases to support clinical management: current perspectives. Pharmgenomics Pers Med 7:275–283
33. Mirnezami R, Nicholson J, Darzi A (2012) Preparing for precision medicine. N Engl J Med 366(6):489–491
34. Ibrahim A et al (2015) Case study for integration of an oncology clinical site in a semantic interoperability solution based on HL7 v3 and SNOMED-CT: data transformation needs. AMIA Jt Summits Transl Sci Proc 2015:71
35. Omidi Y (2011) Smart multifunctional theranostics: simultaneous diagnosis and therapy of cancer. Bioimpacts 1(3):145–147
36. Wang J et al (2014) Smartphone interventions for long-term health management of chronic diseases: an integrative review. Telemed J E Health 20(6):570–583
37. Boulos MNK et al (2011) How smartphones are changing the face of mobile and participatory healthcare: an overview, with example from eCAALYX. Biomed Eng Online 10:24
38. Free C et al (2013), The effectiveness of mobile-health technology-based health behaviour change or disease management interventions for health care consumers: a systematic review. PLoS Med 10(1)
39. Mosa ASM, Yoo I, Sheets L (2012) A systematic review of healthcare applications for smartphones. BMC Med Inform Decis Mak 12:67
40. Pope L, Silva P, Almeyda R 2010 I-phone applications for the modern day otolaryngologist. Clin Otolaryngol 35(4):350–354
41. Pramana G et al (2014) The SmartCAT: an m-health platform for ecological momentary intervention in child anxiety treatment. Telemed J E Health 20(5):419–427
42. Yu F et al (2012) A smartphone application of alcohol resilience treatment for behavioral self-control training. Conf Proc IEEE Eng Med Biol Soc 2012:1976–1979
43. Bhat S et al (2015) Is there a clinical role for smartphone sleep apps? Comparison of sleep cycle detection by a smartphone application to polysomnography. J Clin Sleep Med 11(7):709–715
44. Becker S et al (2015) Demographic and health related data of users of a mobile application to support drug adherence is associated with usage duration and intensity. PLoS One 10(1):e0116980
45. Becker S et al (2013) User profiles of a smartphone application to support drug adherence – experiences from the iNephro project. PLoS One 8(10):e78547
46. Kanawong R et al (2012) Automated tongue feature extraction for ZHENG classification in traditional Chinese medicine. Evid Based Complement Alternat Med 2012:912852
47. Robbins RN et al (2014) A smartphone app to screen for HIV-related neurocognitive impairment. J Mob Technol Med 3(1):23–26

48. Bajaj JS et al (2013) The Stroop smartphone application is a short and valid method to screen for minimal hepatic encephalopathy. Hepatology 58(3):1122–1132
49. Sposaro F, Tyson G (2009) iFall: an android application for fall monitoring and response. Conf Proc IEEE Eng Med Biol Soc 2009:6119–6122
50. Tarbert CM, Livingstone IA, Weir AJ (2014) Assessment of visual impairment in stroke survivors. Conf Proc IEEE Eng Med Biol Soc 2014:2185–2188
51. Park JY et al (2014) Lessons learned from the development of health applications in a tertiary hospital. Telemed J E Health 20(3):215–222
52. Agboola S, Kamdar M (2014) Pain management in cancer patients using a mobile app: study design of a randomized controlled trial. JMIR Res Protoc 3(4):e76
53. Cafazzo JA et al (2015) Usability and feasibility of an mHealth intervention for monitoring and managing pain symptoms in sickle cell disease: the sickle cell disease mobile application to record symptoms via technology (SMART). J Med Internet Res 39(3):162–168
54. Charpentier G et al (2011) The Diabeo software enabling individualized insulin dose adjustments combined with telemedicine support improves HbA1c in poorly controlled type 1 diabetic patients: a 6-month, randomized, open-label, parallel-group, multicenter trial (TeleDiab 1 study). Diabetes Care 34(3):533–539
55. Worringham C, Rojek A, Stewart I (2011) Development and feasibility of a smartphone, ECG and GPS based system for remotely monitoring exercise in cardiac rehabilitation. PLoS One 6(2):e14669
56. Marshall A, Medvedev O, Antonov A (2008) Use of a smartphone for improved self-management of pulmonary rehabilitation. Int J Telemed Appl: p 753064
57. Ryan D et al (2005) Mobile phone technology in the management of asthma. J Telemed Telecare 11(Suppl 1):43–46
58. Atreja A, Khan S (2015) Impact of the mobile Health Promise platform on the quality of care and quality of life in patients with inflammatory bowel disease: study protocol of a pragmatic randomized controlled trial. JMIR Res Protoc 4(1): e23
59. Bangsberg DR, Pan D, Dhall R (2015) A mobile cloud-based Parkinson's disease assessment system for home-based monitoring. J Med Internet Res 3(1):e29
60. Bosl W et al (2013) Scalable decision support at the point of care: a substitutable electronic health record app for monitoring medication adherence. Interact J Med Res 2(2):e13
61. Cho MJ, Sim JL, Hwang SY (2014) Development of smartphone educational application for patients with coronary artery disease. Healthc Inform Res 20(2):117–124
62. Franckle T, Haas D, Mandl KD (2013) App store for EHRs and patients both. AMIA Jt Summits Transl Sci Proc 2013:73
63. Goh G, Tan NC (2015) Short-term trajectories of use of a caloric-monitoring mobile phone app among patients with type 2 diabetes mellitus in a primary care setting. J Med Internet Res 17(2):e33
64. Csernansky JG, Smith MJ (2011) Thought, feeling, and action in real time – monitoring of drug use in schizophrenia. Am J Psychiatry 168(2):120–122
65. Swendsen J, Ben-Zeev D, Granholm E (2011) Real-time electronic ambulatory monitoring of substance use and symptom expression in schizophrenia. Am J Psychiatry 168(2):202–209
66. Sands BE et al (2015) Feasibility of a lifestyle intervention for overweight/obese endometrial and breast cancer survivors using an interactive mobile application. JMIR Res Protoc 137(3):508–515
67. Carter MC et al (2013) Adherence to a smartphone application for weight loss compared to website and paper diary: pilot randomized controlled trial. J Med Internet Res 15(4):e32
68. Casey M et al (2014) Patients' experiences of using a smartphone application to increase physical activity: the SMART MOVE qualitative study in primary care. Br J Gen Pract 64(625):e500–e508
69. Dafli E, Antoniou P (2015) Virtual patients on the semantic Web: a proof-of-application study. J Med Internet Res 17(1):e16

70. Ward MM et al (2003) Participatory patient-physician communication and morbidity in patients with systemic lupus erythematosus. Arthritis Rheum 49(6):810–818
71. Durand MA et al (2014) Do interventions designed to support shared decision-making reduce health inequalities? A systematic review and meta-analysis. PLoS One 9(4):e94670
72. Goddu AP, Raffel KE, Peek ME (2015) A story of change: the influence of narrative on African-Americans with diabetes. Patient Educ Couns 98(8):1017–1024
73. Lejbkowicz I, Caspi O, Miller A (2012) Participatory medicine and patient empowerment towards personalized healthcare in multiple sclerosis. Expert Rev Neurother 12(3):343–352
74. Majmudar MD, Colucci LA, Landman AB (2015) The quantified patient of the future: opportunities and challenges. Healthc (Amst) 3(3):153–156
75. Call J et al (2012) Survival of gastrointestinal stromal tumor patients in the imatinib era: life raft group observational registry. BMC Cancer 12:90
76. Kear T, Harrington M, Bhattacharya A (2015) Partnering with patients using social media to develop a hypertension management instrument. J Am Soc Hypertens 9(9):725–734
77. Hood L, Auffray C (2013) Participatory medicine: a driving force for revolutionizing healthcare. Genome Med 5(12):110
78. Palmer JE (2012) Genetic gatekeepers: regulating direct-to-consumer genomic services in an era of participatory medicine. Food Drug Law J 67(4):475–524. iii
79. Reeves S et al (2017) Interprofessional collaboration to improve professional practice and healthcare outcomes. Cochrane Database Syst Rev, CD000072.pub3
80. Jain M et al (2006) Decline in ICU adverse events, nosocomial infections and cost through a quality improvement initiative focusing on teamwork and culture change. Qual Saf Health Care 15(4):235–239
81. Almalki M, Gray K, Sanchez FM (2015) The use of self-quantification systems for personal health information: big data management activities and prospects. Health Inf Sci Syst 3(Suppl 1. HISA Big Data in Biomedicine and Healthcare 2013 Con):S1
82. Kuziemsky C et al (2014) A framework for incorporating patient preferences to deliver participatory medicine via interdisciplinary healthcare teams. AMIA Annu Symp Proc 2014:835–844
83. Bredfeldt C et al (2015) Patient reported outcomes for diabetic peripheral neuropathy. J Diabetes Complications 29(8):1112–1118
84. Frost J et al (2011) Patient-reported outcomes as a source of evidence in off-label prescribing: analysis of data from PatientsLikeMe. J Med Internet Res 13(1):e6
85. Norris K (2014) Lung cancer patient advocacy and participatory medicine. Genome Med 6(1):7
86. Charani E et al (2014) Do smartphone applications in healthcare require a governance and legal framework? It depends on the application! BMC Med 12:29

# Chapter 2
# Physiological Informatics: Collection and Analyses of Data from Wearable Sensors and Smartphone for Healthcare

Jinwei Bai, Li Shen, Huimin Sun, and Bairong Shen

**Abstract** Physiological data from wearable sensors and smartphone are accumulating rapidly, and this provides us the chance to collect dynamic and personalized information as phenotype to be integrated to genotype for the holistic understanding of complex diseases. This integration can be applied to early prediction and prevention of disease, therefore promoting the shifting of disease care tradition to the healthcare paradigm. In this chapter, we summarize the physiological signals which can be detected by wearable sensors, the sharing of the physiological big data, and the mining methods for the discovery of disease-associated patterns for personalized diagnosis and treatment. We discuss the challenges of physiological informatics about the storage, the standardization, the analyses, and the applications of the physiological data from the wearable sensors and smartphone. At last, we present our perspectives on the models for disentangling the complex relationship between early disease prediction and the mining of physiological phenotype data.

**Keywords** Wearable sensors • Smartphone • Physiological informatics • Participatory medicine • Data mining for healthcare

---

J. Bai
Digital Department of Library, Soochow University, Suzhou 215006, China

L. Shen • B. Shen (✉)
Center for Systems Biology, Soochow University, No.1 Shizi Street, Suzhou, Jiangsu 215006, China
e-mail: bairong.shen@suda.edu.cn

H. Sun
Collaborative Innovation Center of Sustainable Forestry in Southern China of Jiangsu Province, Nanjing Forestry University, Nanjing 210037, China

© Springer Nature Singapore Pte Ltd. 2017
B. Shen (ed.), *Healthcare and Big Data Management*, Advances in Experimental Medicine and Biology 1028, DOI 10.1007/978-981-10-6041-0_2

## 2.1 Introduction

### 2.1.1 From Disease Care to Healthcare: The Coming of Age

Based on the theory of Kondratiev's long economic cycle, the sixth economic wave is coming with the technological innovations in the fields of psychosocial health and biotechnology [1]. Furthermore, the scientific paradigm and social requirements are also shifting toward the healthcare. The healthcare is becoming the most important issue in our life driven by the three forces shown in Fig. 2.1. In the near future, we are facing the limited medical resources with the arrival of the aging era and the lack of labors in the markets. To reduce the social and family burden for disease care, it is necessary to move from clinical management to early disease prevention and healthcare. As proposed by Leroy Hood [2, 3], the P4 medicine (i.e., predictive, preventive, personalized, and participatory medicine) will be the new paradigm for disease prevention and treatment. The efficient prediction and early prevention of complex diseases can be very helpful, and it could reduce about 75% of the disease care cost in the USA [4].

Among many outstanding technical innovations in the sixth Kondratiev's economic cycle, the application of wearable sensors and smartphone to the health/disease monitoring is becoming widespread. These smart wearable sensors could be watches, caps, clothing, shoes, patches, tattoos, body ornaments, etc. [5]. These wearable sensors are easy to use, and many kinds of physiological data such as the posture and gait; the skeletal muscle movement (electromyogram or electromyography, EMG); heart rate and tracking (electrocardiogram, ECG); blood pressure, temperature, sleeping, and brain activity (electroencephalograms, EEG); skin hydration, blood oxygen level, medication ingestion, eye moving tracking (electrooculogram, EOG); and respiration signals could be detected continuously and real-time monitored remotely. These physiological signals from smart wearable sensors could be collected, stored, and analyzed with smartphone and cloud computing for the further applications to the disease management and healthcare.

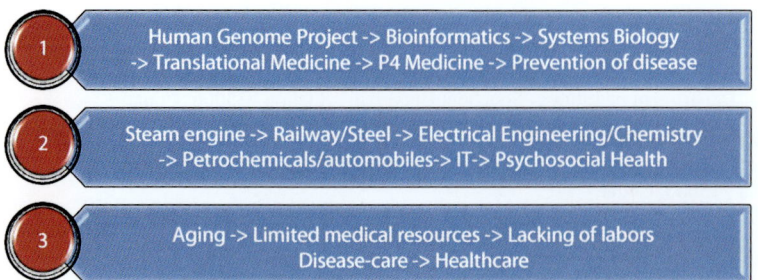

**Fig. 2.1** Three driven forces for healthcare development. (*1*) Scientific paradigm shifting; (*2*) the economic long wave changing; (*3*) the changing age structure

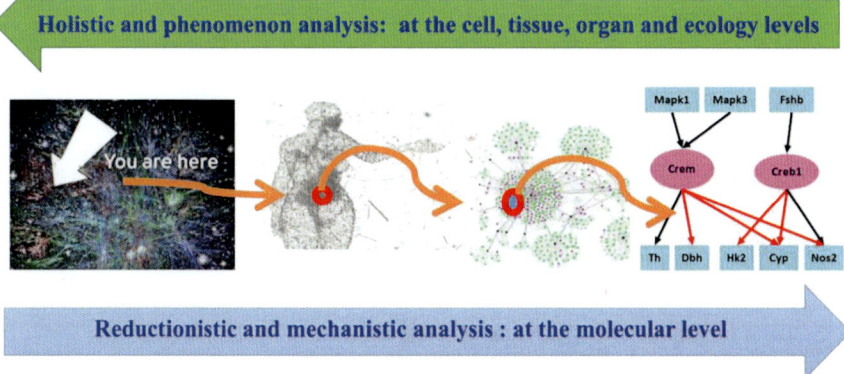

**Fig. 2.2** Reductionism vs. holism in biomedical researches

## 2.1.2 Physiological Signals as Phenotyping Data

As we know, with the progress of the human genome project and the next-generation sequencing technologies, the detection and collection of genotyping data become cheaper and cheaper, and the reconstruction of gene network for a concrete disease becomes reality. Considering the genotype-phenotype relationship, the phenotyping nowadays is still coarse grained, and we do not have enough phenotyping data for a concrete disease. Therefore the paired genotype-phenotype data are mostly lacking for personalized treatment of complex diseases in recent years. The studies in molecular level are often based on the reductionism hypothesis. The molecular network reconstruction or the systems-level analysis is seldom applied at the tissue or organ levels and no mention the individual whole-body level, as presented in Fig. 2.2.

We nowadays can obtain the static genetic structure of a patient with payable cost. We can also measure the gene expression (transcriptomic), the proteomic, and the metabolomic data. We can even obtain the Omics data for a single cell in the human cancer tissues. But it would be more than difficult to study and model the interactions between cells, tissues, and organs. The complex diseases are caused by the interaction between the human body and the environment dynamically. The study of disease only at the molecular level may not be enough to understand the "whole elephant" of the human body as presented in Fig. 2.2. We need to integrate the information from gene or molecular level to signals from the high levels like cell, tissue/organ, individual, and ecology scales. The physiological signals reflect the dynamic interactions between the individuals and their environments. These signals obtained from wearable sensors provide us the chance to collect fine, dynamic, and personalized physiological phenotype data in real time. This can help us to monitor the personalized health state in precision.

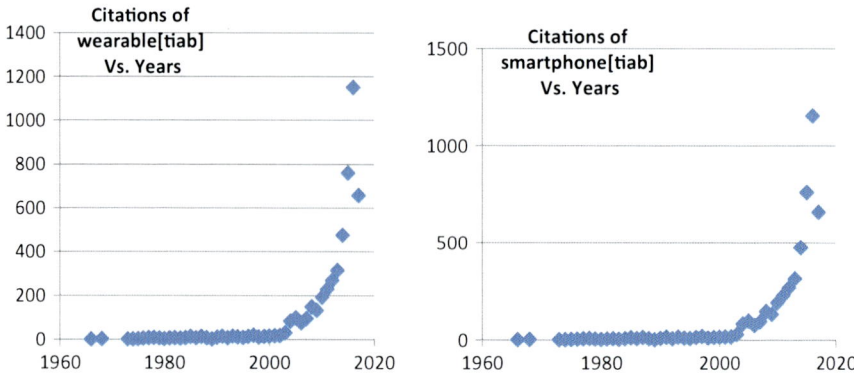

**Fig. 2.3** The PubMed citations for "wearable [tiab]" and "smartphone [tiab]"

By combining the use of wearable sensors with the smartphone and cloud computing, it will be possible to store the physiological data in cloud. These data can be analyzed and linked to a patient's medical doctors, family members, etc.; we can expect that the disease risk-associated signals will be detected, analyzed, and connected to immediate actions. We can then efficiently predict and prevent the complex diseases, and we will at last transfer the disease care tradition to the healthcare paradigm. But before that, we have many challenges in transferring the dynamic and real-time big physiological data to healthcare wisdoms.

### 2.1.3 The Challenge for the Mining of Physiological Data from Wearable Sensors

As shown in Fig. 2.3, the numbers of publications on wearable sensors and smartphone are increasing year by year in the past decades. We will have more and more physiological phenotype data collected and algorithms developed to mine these data that will be demanded specific to different healthcare questions. In this chapter, we will discuss the present state of physiological informatics and the future challenges for the mining of the data.

The pipeline of collection of physiological data from wearable sensors and mining these data for healthcare wisdoms is scratched in Fig. 2.4. The questions arising from the analyses of the physiological data include the following: (1) What kind of data could be detected by the wearable sensors? (2) How to share and analyze the detected physiological profiles? (3) How to use the physiological features for healthcare and prediction of diseases? (4) What are the challenges for the data analyses and applications of the physiological data? We will review on the first three questions in Sects. 2 and 3, and the last question will be discussed in Sect. 4.

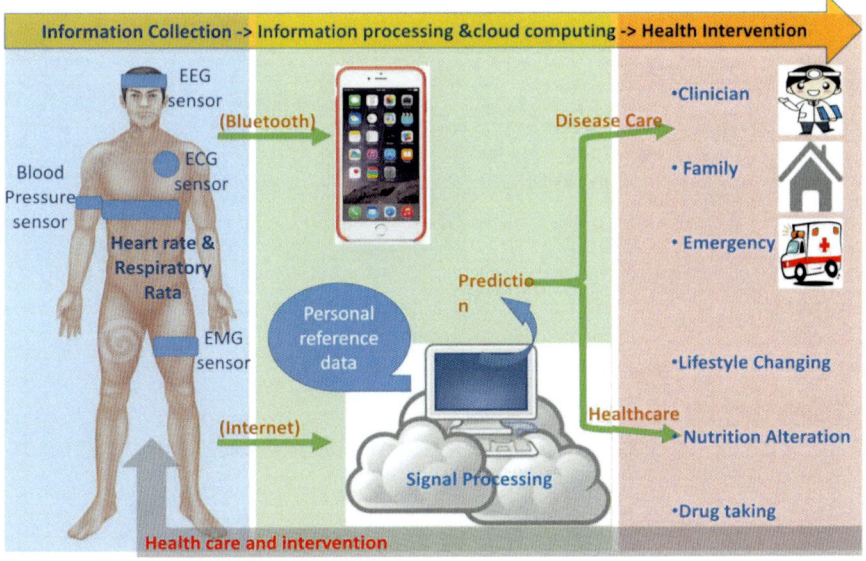

**Fig. 2.4** The pipeline of physiological data collection to healthcare wisdoms

## 2.2 The Collection of the Big Physiological Data from the Wearable Sensors

### 2.2.1 The Traditional Methods for Physiological Data Collection and Analyses

Human health states are always associated with some basic physiological indices such as body temperature, blood pressure, heart rate, pulse, respiratory rate, ECG, EEG, EOG, EMG, etc. Traditionally, these data are often checked and collected periodically when the patients visit hospitals. Special conditions and devices are often needed. The collection of the dynamic physiological data is often time-consuming, labor-intensive, and costly. The analyses of these data are often statistically averaged to identify the patterns at the population level, as described in Fig. 2.5.

The advantage of this traditional method is that it can identify general physiological patterns at population level and it could provide facts, evidence, and reference for the policy making and disease screening. But it will be not precise when applied to individuals since the averaged indices are not personalized.

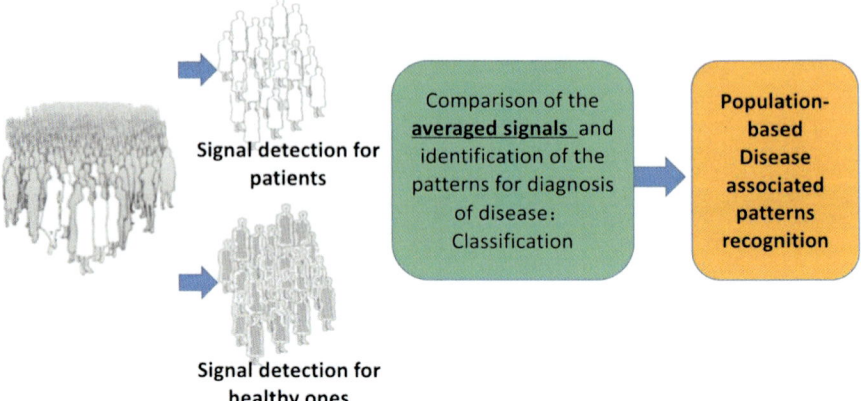

**Fig. 2.5** Cohort data collection and statistical analysis to identify healthcare-associated factors

### 2.2.2 The Longitudinal Self-Measurements by Wearable Sensors and the Smartphone-Based Cloud Computing for Data Storage, Extraction, and Analysis

In the past two decades, the popularization of wearable sensors and smartphone is changing of our life rapidly. The whole-body monitoring by wearable sensors is becoming more and more practical. We nowadays have the potential to monitor the whole body's physiological signals for disease prevention from head to foot. In the head part, we can use wearable sensors to detect the brain activity by electroencephalograms (EEG) for monitoring of epilepsy, fatigue, mental stress, anxiety states, etc. [6–9]. The eye's movement could be measured with electrooculogram (EOG) to monitor older adults and patients with Parkinson's disease. The facial expression could be detected for emotional recognition [10]; the gait and balance and the fall risk could be detected to monitor the foot part. The other detectable physiological signals of the human body include the electrocardiogram (ECG) for the heart rhythm, electromyogram (EMG) for skeletal muscle movement, etc. All these physiological signals could be detected and applied to the monitoring of a wide spectrum of diseases.

Comparing to the traditional methods, the wearable sensors are cheap and convenient. It could be applied to both patients and health individuals, and the data could be collected in real time, dynamic, and personalized as shown in Fig. 2.6. The volume of these physiological data will be accumulated very fast. With different sensors, we will have diverse data formats, and the velocity of the data generating will be also high. So the physiological data from the wearable sensors possess distinct characterization of big data. All the tools and methods for big data management therefore could be applied here for the big physiological data storage, extraction, and analysis.

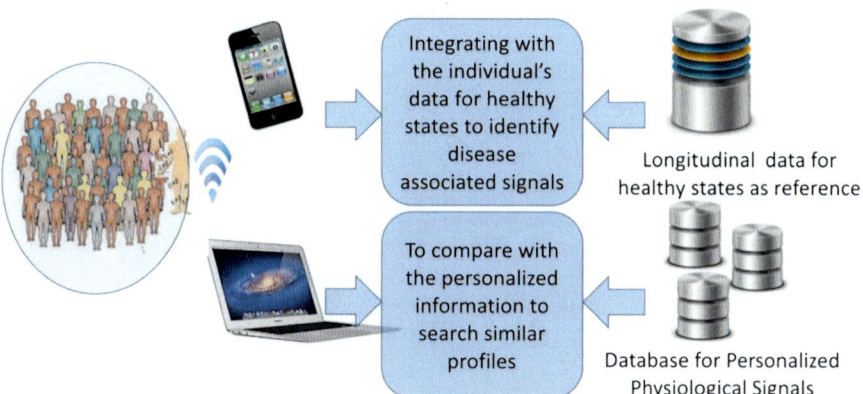

Fig. 2.6 Real-time and personalized analyses of longitudinal measurements by wearable sensors

Thanks to the cloud computing technologies, the big physiological data could be stored, extracted, or even get analyzed results from the cloud with smartphone linked to the internet. The cloud computing providers can offer a user different service including software tools (SaaS, software as a service), platforms (PaaS), and infrastructure (IaaS). As displayed in Fig. 2.6, the detected personalized data could be collected, accumulated, and stored in cloud databases as reference data. A personalized physiological data could be compared to the reference data to find similar profiles and then screen better treatment strategies. The individual's previous health data or disease data could also be applied to the diagnosis and treatment of complex diseases. This smartphone-based cloud service was reported to be used to manage type 1 diabetes (T1D) and chronic obstructive pulmonary disease (COPD) patients with comorbidities [11, 12].

## 2.3 Data Mining of the Physiological Data and the Challenges

### 2.3.1 Data Standardization and the Privacy of Personal Physiological Information

Data standardization is important to the exchange and sharing of big data between researchers, companies, organizations, and other data users. The physiological data could be generated from different resources with various structures, formats, or terminologies. Ontology-based standardization is the first step to make these diverse data sharable and reusable (see Fig. 2.7).

Ontology is a knowledge framework with a controlled vocabulary and defined relationship between them to be used in a subject or a domain for identification of themes and patterns in a given data set [13]. At present, several ontologies are

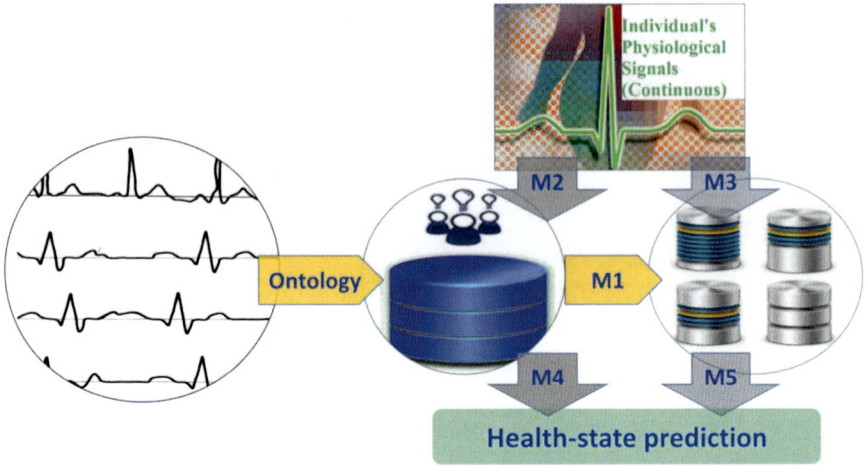

**Fig. 2.7** Models for mining physiological data

developed and applied to physiological data and can be extended to the storage and analyses of the data collected from different wearable sensors. Some of the developed ontologies for the standardization of physiological data are listed in Table 2.1. More ontologies are still needed for the diversity of physiological data from smart wearable sensors, including the basic and high-level physiological information like the pulse, the blood pressure, and the dynamic patterns of the personalized signals. The ontologies will not be only useful for the data sharing as they could be very powerful tools for the annotation and explanation of the data and the ontological functional analysis, then enabling and accelerating the researches [14].

For the data sharing, the privacy of the personalized data is another important issue needed to be resolved before the data distributed to public. Several concepts and frameworks are proposed for the privacy preserving of data from wearable sensors. The stringent CIA (confidentiality, integrity, and availability) and Health Insurance Portability and Accountability Act (HIPAA) principles are suggested to follow for the information security [23–27].

## 2.3.2 Databases and Methods for the Mining of Physiological Signals

After the data standardization, different databases specific to various physiological data are then needed. In the past, some physiological databases have been built for public accessing. Comparing to the databases at gene level, physiological phenotype database are still demanded. One of the most comprehensive databases is the PhysioBank database from PhysioNet resource, which includes physiological signals like ECG, interbeat interval, gait and balance, neuroelectric and myoelectric,

**Table 2.1** Ontologies developed and applied in physiological data

| Ontologies | Description and availability | References |
|---|---|---|
| Ion Channel ElectroPhysiology Ontology (ICEPO) | Ontological representation for extracting quantitative information from text | [15] |
| | http://openbionlp.org/mutd/supplementarydata/ICEPO/ICEPO.owl | |
| OntoVIP | The annotation of the models used in medical image simulation | [16] |
| | http://bioportal.bioontology.org/ontologies/OntoVIP | |
| Ontology of Physics for Biology (OPB) | The representations of the thermodynamics and dynamics of physiological processes | [17] |
| | http://bioportal.bioontology.org/ontologies/OPB | |
| Epilepsy and Seizure Ontology (EpSO) | A suite of informatics tools for representation and study of epilepsy | [18] |
| | http://prism.case.edu/prism/index.php/EpilepsyOntology | |
| Human Physiology Simulation Ontology (HuPSON) | A framework for biomedical physiological simulation | [19] |
| | http://bishop.scai.fraunhofer.de/scaiview/ | |
| Cellular Phenotype Ontology (CPO) | Ontology for characterization of cell morphology and physiological phenotypes | [20] |
| | http://cell-phenotype.googlecode.com | |
| ECG Ontology | Ontology based on the HL7 standard | [21] |
| Hierarchical Event Descriptors (HED) | Semi-structured tagging for EEG http://sccn.ucsd.edu/eeglab | [22] |

image, etc. In the PhysioNet webpage, you can also download software tools for viewing and analyzing of physiologic signals. Many associated physiological databases are also collected there, including MIT-BIH Arrhythmia, European ST-T, Long-Term ST, MIT-BIH Noise Stress Test, Creighton University Ventricular Tachyarrhythmia, MIT-BIH Atrial Fibrillation, and MIT-BIH Supraventricular Arrhythmia and Normal Sinus Rhythm. Other data resources and databases are also linked and updated in the following webpage: https://physionet.org/other-links.shtml or https://physionet.org/physiobank/other.shtml. Some of the databases are listed in Table 2.2, although the list is absolutely far from comprehensive.

As indicated in Fig. 2.7, one of the big challenges to understand the complex physiological signals for healthcare or disease management is the building of specific data analysis models, such as (1) clustering or classifying individuals' physiological data to distinct groups based on their profile similarities (M1 in Fig. 2.7), (2) comparing an individual's physiological feature to the population samples to identify similar health/disease profiles (M2), (3) classifying the individual's physiological signals from known groups (M3), and (4) optimizing a score function or model to classify a given physiological profile to predict the health state (M4 and M5).

**Table 2.2** Databases for physiological signals

| Database | Description and availability | References |
|---|---|---|
| AHA ECG DB | An ECG database by the American Heart Association | [28] |
| | https://www.ecri.org/components/Pages/AHA_ECG_USB.aspx | |
| DEAP | ECG and peripheral physiological signals | [29] |
| | http://www.eecs.qmul.ac.uk/mmv/datasets/deap/ | |
| ECG-ViEW II | Electrocardiogram vigilance with electronic data warehouse II | [30] |
| | http://www.ecgview.org | |
| Ann Arbor Electrogram Libraries | Annotated recordings of intracardiac electrograms | [31] |
| | http://electrogram.com/ | |
| MIT PhysioNet | It offers free physiologic data (PhysioBank) and software tools (PhysioToolkit) | [32–36] |
| | http://physionet.org/ | |
| TUH EEG Corpus | The Temple University Hospital EEG Data Corpus | [37] |
| | https://www.isip.piconepress.com/projects/tuh_eeg/html/downloads.shtml | |
| The DREAMS Sleep Spindles Database | Data were collected during the DREAMS project | [38, 39] |
| | http://www.tcts.fpms.ac.be/~devuyst/#Databases | |
| MAREA (Movement Analysis in Real-world Environments using Accelerometers) Gait Database | It includes information about gait activities in different real-world environments | [40] |
| | http://islab.hh.se/mediawiki/Gait_database | |
| Gait MoCap database | Data for gait cycles and bone rotation | [41] |
| | https://gait.fi.muni.cz/ | |

In the previous 20 years, many models are proposed to discover patterns in the diverse physiological data. Some of the models are selected and listed in Table 2.3 with examples and description. Most of the models are feature-based pattern detection, scoring, or classification. Point scoring methods are often empirical with clinicians' intuition and experience. Motif and signal transformation methods are often robust to noise. Signal transformation methods are diverse, and it could be Fourier, wavelet, bilinear, Hilbert-Huang, Stockwell, and Laplacian transformation [47, 49–54], considering the different characterization of the physiological information. Entropy methods are often combined to signal transformation for better

**Table 2.3** Methods applied for the analyses of physiological signals

| Method | Example | Description | References |
|---|---|---|---|
| Feature extraction | Sleep stage classification | The features used for the classification are Hjorth features, wavelet transformation, and symbolic representation | [42] |
| Point scoring | Diagnosis of left ventricular hypertrophy (LVH) | QRS amplitude/axis/duration, ST-T segment, left atrial involvement, and intrinsicoid deflection were given different point scores for the evaluation of LVH risk | [43, 44] |
| Motif | Anomaly detection in ECG | Motif discovery based on expert knowledge is robust and accurate for anomaly detection | [45] |
| Entropy | Elucidating the therapeutic effects of antipsychotics | Multiscale entropy (MSE) is examined to identify abnormal dynamical EEG signal complexity and to elucidate antipsychotics' therapeutic mechanisms | [46] |
| Signal transformation | Hypoxia ischemia (HI) EEG detecting | Robust wavelet transformation was applied to automatically detect sharp waves in the HI-EEG | [47] |
| Machine learning | Quantifying the depth of anesthesia (DoA) | A novel method (HDoA) was introduced to accurately quantify DoA by analyzing EEG using hidden Markov model | [48] |

classification of the signals. With more data collected and available, machine learning or deep learning methods could be widely applied to the physiological data analysis to identify the hidden structures for precise prediction, classification, and pattern recognition. The widely used machine learning methods include hidden Markov model, artificial neural network, support vector machine, and so on. These methods require enough data for training and testing to prevent over-fitting, which is the disadvantage of this method.

### 2.3.3 ECG Data Analysis as a Case Study

Since the physiological signals can reflect the whole body's state. The human body is a complex system which is robust to the environment disturbing. The changes of physiological indices are often related to the health state alteration. In the ancient, the pulse, body temperature, and other basic physiological signals are the basic evidence used by traditional Chinese medical doctors for diagnosis of diseases. Even at present, the physiological signals are still the important indicators of health state. Especially the personalized, dynamic, and real-time information detected by wearable sensors are the firsthand evidence for the diagnosis and disease management. We here take ECG as an example to reveal the importance of the

**Table 2.4** ECG profile analysis and applications

| Application | Methods or data | Models or results | References |
|---|---|---|---|
| Atrial fibrillation (AF) | Smartphone-based wireless single-lead ECG data were collected for 13,122 Hong Kong citizens | A significant proportion of AF patients were newly diagnosed by using multivariable logistic regression model | [55] |
| Cardiovascular disease (CVD) | Data are collected from an energy-efficient ECG signal processor | A forward search algorithm is implemented in an integrated circuit for diagnosis of CVD on smartphone | [56] |
| Arrhythmia detection | ECG is collected by the patient-worn ECG sensor and remote monitoring station | The heartbeats were detected by Pan-Tompkins algorithm and then classified by a decision tree | [57] |
| ST-segment elevation myocardial infarction (STEMI) | ECG signals from smartphone were taken to compare with the traditional 12-lead ECG for evaluation of STEMI | The ECG signal from smartphone shows excellent correlation with the gold standard 12-lead ECG | [58] |
| Myocardial ischemia (MI) | Real-time ECG monitoring was applied to a 42-year-old radical prostatectomy patient with previous MI. The monitoring was linked to the clinician's smartphone | This study indicates that the remote ECG monitoring has the potential to detect myocardial ischemic and to prevent postoperative MI | [59] |
| Fall risk detection | Heart rate variability (HRV) was extracted from ECG to assess the fall risk | Model for classification and regression was developed, and the performance is satisfactory | [60] |
| Palpitation | Athletes with palpitation were monitored by ECG, and the signals were sent to clinicians for diagnosis | The ECG monitoring can enhance the evaluation of symptomatic athletes | [61] |
| Sleep-wake discrimination | A new adaptive classification algorithm was embedded on a wearable system to process the ECG and respiratory effort signals | Sleep/wake could be precisely classified without the intervention of experts or off-line calibration by the suggested automatic adaptation method | [62] |

physiological signals for the disease management or healthcare. ECG is one of the widely used physiological signals for the diagnosis and intervention of diseases.

In Table 2.4, we list many of the applications which take the ECG profiles as the evidence to diagnose the heart diseases and other abnormalities such as atrial fibrillation, arrhythmia detection, ST-segment elevation myocardial infarction (STEMI), myocardial ischemia (MI), fall risk detection, palpitation, etc.

As we know, heart disease is the disease with the highest mortality rate worldwide. If we can take use of the wearable sensor to monitor the individual health state to alert the clinicians or family members, it will reduce the mortality rate of the heart disease patients. Until now, we have many static ECG profiles characterized and related to different symptoms. With the dynamic and real-time personalized data available, we can expect that more useful dynamic ECG patterns can be identified for the precise prediction and prevention of heart diseases.

## 2.4 Perspectives on the Integrative Analyses of the Physiological Data

In the big data era, big physiological data is only one part of the big biomedical data. The human body is a very complex, dynamic, and personalized system. It is still a big challenge to study the robust, evolutionary system. We now have the chance to take use of all related biomedical big data for the integration of the information from different "blind men" for the scratch of the big "elephant."

Within this section, we'll discuss the cross-level integrative analysis of the physiological data with other data such as the gene expression data, the imaging data, the public health data, and so on. We will explain the systems thinking, the patient-centered (or person-centered) participatory medicine, the biopsychosocial model of health considering the systems evolution and non-drug therapy, etc.

### 2.4.1 Systems Thinking and the Integrative Analysis of the Physiological Data

It is well known that complex disease is heterogeneous. The disease phenotypes are often generated by the interaction between genes, lifestyle, and environments. We previously demonstrated that we cannot identify common set of genes associated to a complex disease, but we may identify common patterns at systems level, such as pathways and networks [63–69]. Therefore systems thinking is essential for understanding the mechanisms and the heterogeneity of a complex disease. Complex disease itself is also robust, a disease can always develop a specific strategy to escape from the immune systems' checking and surveillance, and drug resistance is often presented in the late stage of disease evolution [70]. From a systems biological view, the disease phenotype could be caused by an altered gene network that could be reconstructed by integration analysis of different Omics data, which is typically called reverse engineering of gene interaction. Taking the physiological signals as phenotype, we therefore could establish links between the gene network and the physiological signals. In Table 2.5, we list some examples of integrative analysis of physiological data with genetic data, imaging data, individual clinical data, and public health data. The data and information from the four levels, i.e., molecule, cell/tissue, and individual to population, can be integrated to understand

**Table 2.5** Integrative analysis of physiological data

| Physiological data+ | Example | References |
|---|---|---|
| Genetic data | The ECG T-wave patterns are found associated with genetic loci for the long QT syndrome | [71] |
| | QRS fragmentation and S-wave features of ECG profile can be used to predict arrhythmias in SCN5A D1275N mutation carriers | [72] |
| Imaging features | To identify myoclonic-astatic epilepsy associated with neural networks by integration of EEG and fMRI information | [73] |
| | Integration of fMRI and ECG data to identify brain structures for pain-related autonomic changes | [74] |
| Clinical symptoms | An integrating study of EEG abnormalities and depressive symptom outcomes | [75] |
| | An integrative study of EEG features and symptom severity in PTSD (post-traumatic stress disorder) | [76] |
| Public health study | Population-level identification of genetic basis for EEG alpha | [77] |
| | A multicenter study indicates that the quantitative EEG (QEEG) could be a diagnostic marker for dementia with Lewy bodies | [78] |

the human body system, to understand the genotype-phenotype relationship, and to investigate the molecular mechanisms relevant to the relationship.

Based on the information theory, with more information or knowledge provided for a complex system, we should have clearer picture for the complex systems. The cross-level integrative analysis becomes more and more popular with the data available. To collect the gene-level information is becoming cheaper and cheaper. The paired genotype-phenotype data are really demanded for precision medicine and healthcare.

### 2.4.2 The Paradigm for the Participatory Medicine

For the collection of paired data at different levels for the systems investigation of complex disease or healthcare, the fourth P of P4 medicine, i.e., the participatory medicine, is now becoming practical with the technology developing of wearable sensors, smartphone, and wireless internet/energy. Participatory medicine is a new paradigm for disease management or healthcare, which is based on the patient-centered (or person-centered) network linked to clinicians, patients' family member and friends, etc. Within the network, the distance between individuals is shortened base on "six degrees of separation" theory [79]. The patients can provide and obtain data, information, knowledge, or experience from the network in a convenient way. The informaticians behind the internet/network can provide the users analytic tools and services for the prediction, prevention, and understanding of their health state. PatientsLikeMe is a novel model and platform for real-time research and participatory medicine paradigm [80]. It can ensure the data scientists and researchers

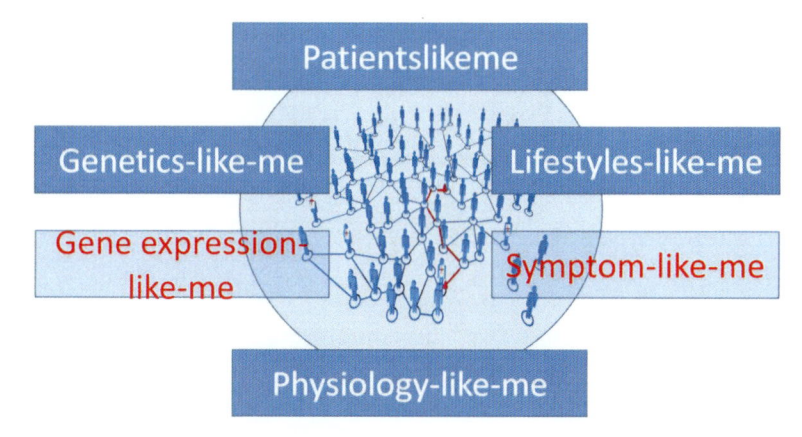

**Fig. 2.8** Physiology-Like-Me and participatory medicine

enough data resource for the disease management and healthcare modeling and simulations.

With the application of wearable sensors and smartphone, the personalized physiological data and profiles can be obtained in real time; the Physiology-Like-Me will be popularized and extended as one of the PatientsLikeMe model in the future. The same will happen to lifestyles, gene expression, genetics and symptoms [81], etc., as shown in Fig. 2.8.

## 2.4.3 Healthcare Intervention with Evolutionary Thinking and Non-drug Therapy

Under the systems medicine or systems healthcare paradigm, healthcare intervention by lifestyle changing will be one of the main pathways to improve the health or prevention of chronic diseases. The evolutionary study of disease or heath can be emerged with the time series or dynamic data available. The evolutionary trajectories of complex disease could be inferred reasonably, and the cross-level integrative analyses with the evolutionary ideas make the data analysis more challenging and more fascinating (see Fig. 2.9).

George L. Engel's biopsychosocial model is an integrative model which emphasizes the three factors, i.e., the biological, psychological, and social factors for the medicine and healthcare [82].

The function of non-drug therapy and healthcare should be taken into account for the health promotion, and the related research will be promising to the disease prevention. As shown in Fig. 2.10, many factors are interacted with each other to affect our health state. NIH recently initiated a project to explore the relationship between music and the mind in order to promote the research on music therapy [83]; non-drug therapy including reading and artistic activities must have effects on our

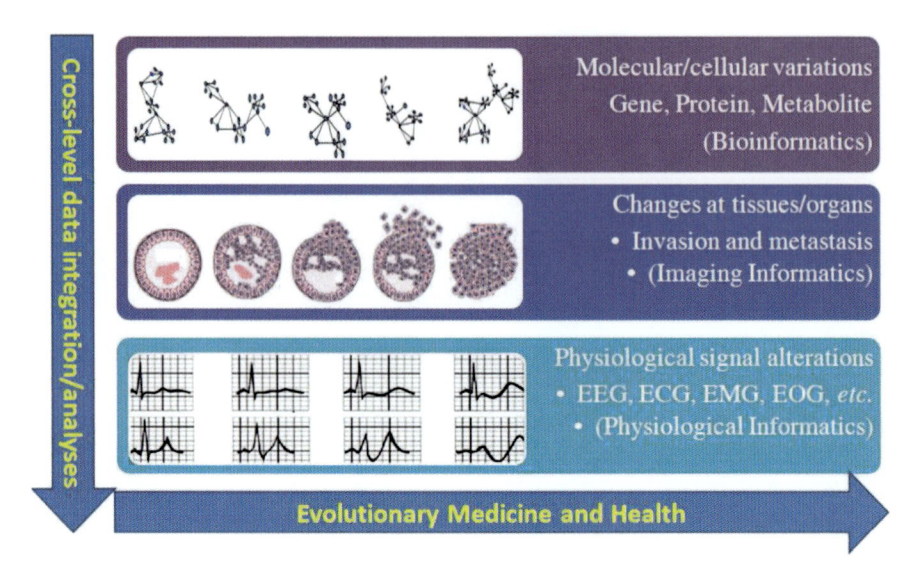

**Fig. 2.9** Cross-level data integration and analyses based on the evolutionary medicine and health paradigm

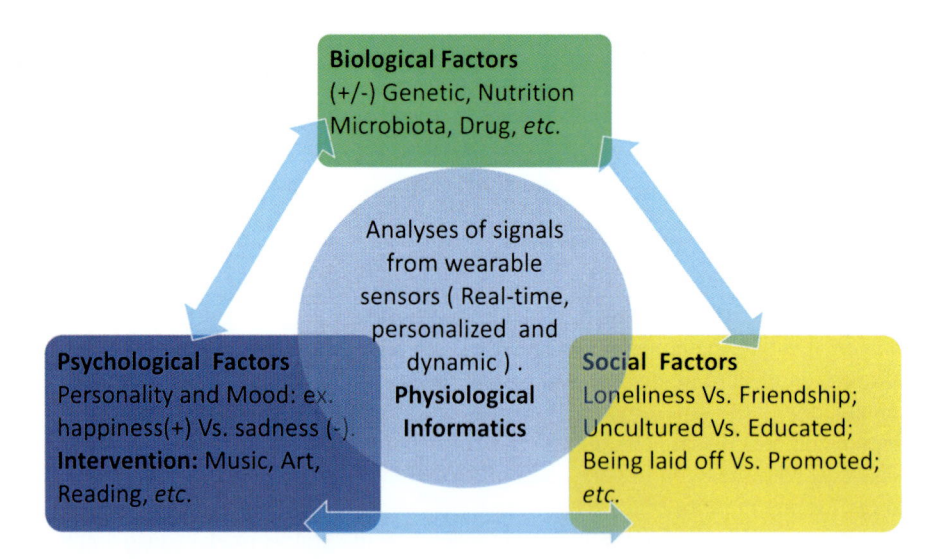

**Fig. 2.10** The dynamic interaction between biopsychosocial factors and their relationship with the physiological signals

health state. Thus considering these factors in our healthcare and intervention will be widely applied.

**Acknowledgments** This study was supported by the National Natural Science Foundation of China (NSFC) (grant nos. 31670851, 31470821, and 91530320) and National Key R&D programs of China (2016YFC1306605).

# References

1. Nefiodow L, Nefiodow S (2014) The sixth Kondratieff: a new long wave in the global economy. ISBN 978-1-4961-4038-8. Charleston
2. Hood L (2008) A personal journey of discovery: developing technology and changing biology. Annu Rev Anal Chem (Palo Alto, Calif) 1:1–43
3. Auffray C, Charron D, Hood L (2010) Predictive, preventive, personalized and participatory medicine: back to the future. Genome Med 2(8):57
4. Flores M, Glusman G, Brogaard K, Price ND, Hood L (2013) P4 medicine: how systems medicine will transform the healthcare sector and society. Personalized Med 10(6):565–576
5. Elenko E, Underwood L, Zohar D (2015) Defining digital medicine. Nat Biotechnol 33(5):456–461
6. McKenzie ED, Lim AS, Leung EC, Cole AJ, Lam AD, Eloyan A, Nirola DK, Tshering L, Thibert R, Garcia RZ et al (2017) Validation of a smartphone-based EEG among people with epilepsy: a prospective study. Sci Rep 7:45567
7. Hu B, Peng H, Zhao Q, Hu B, Majoe D, Zheng F, Moore P (2015) Signal quality assessment model for wearable EEG sensor on prediction of mental stress. IEEE Trans Nanobioscience 14(5):553–561
8. Zhang X, Li J, Liu Y, Zhang Z, Wang Z, Luo D, Zhou X, Zhu M, Salman W, Hu G et al (2017) Design of a fatigue detection system for high-speed trains based on driver vigilance using a wireless wearable EEG. Sensors (Basel) 17(3):E486
9. Asakawa T, Muramatsu A, Hayashi T, Urata T, Taya M, Mizuno-Matsumoto Y (2014) Comparison of EEG propagation speeds under emotional stimuli on smartphone between the different anxiety states. Front Hum Neurosci 8:1006
10. Jangho K, Da-Hye K, Wanjoo P, Laehyun K (2016) A wearable device for emotional recognition using facial expression and physiological response. Conf Proc IEEE Eng Med Biol Soc 2016:5765–5768
11. Baskaran V, Prescod F, Dong L (2015) A smartphone-based cloud computing tool for managing type 1 diabetes in Ontarians. Can J Diabetes 39(3):200–203
12. Chouvarda I, Philip NY, Natsiavas P, Kilintzis V, Sobnath D, Kayyali R, Henriques J, Paiva RP, Raptopoulos A, Chetelat O et al (2014) WELCOME – innovative integrated care platform using wearable sensing and smart cloud computing for COPD patients with comorbidities. Conf Proc IEEE Eng Med Biol Soc 2014:3180–3183
13. Yu C, Shen B (2016) XML, ontologies, and their clinical applications. Adv Exp Med Biol 939:259–287
14. Rubin DL, Shah NH, Noy NF (2008) Biomedical ontologies: a functional perspective. Brief Bioinform 9(1):75–90
15. Elayavilli RK, Liu H (2016) Ion Channel Electro Physiology Ontology (ICEPO) – a case study of text mining assisted ontology development. AMIA Joint Summits Transl Sci Proc AMIA Joint Summits Transl Sci 2016:42–51
16. Gibaud B, Forestier G, Benoit-Cattin H, Cervenansky F, Clarysse P, Friboulet D, Gaignard A, Hugonnard P, Lartizien C, Liebgott H et al (2014) OntoVIP: an ontology for the annotation of object models used for medical image simulation. J Biomed Inform 52:279–292
17. Cook DL, Neal ML, Bookstein FL, Gennari JH (2013) Ontology of physics for biology: representing physical dependencies as a basis for biological processes. J Biomed Semant 4(1):41

18. Sahoo SS, Lhatoo SD, Gupta DK, Cui L, Zhao M, Jayapandian C, Bozorgi A, Zhang GQ (2014) Epilepsy and seizure ontology: towards an epilepsy informatics infrastructure for clinical research and patient care. J Am Med Inform Assoc AMIA 21(1):82–89
19. Gundel M, Younesi E, Malhotra A, Wang J, Li H, Zhang B, de Bono B, Mevissen HT, Hofmann-Apitius M (2013) HuPSON: the human physiology simulation ontology. J Biomed Semant 4(1):35
20. Hoehndorf R, Harris MA, Herre H, Rustici G, Gkoutos GV (2012) Semantic integration of physiology phenotypes with an application to the cellular phenotype ontology. Bioinforma (Oxford, England) 28(13):1783–1789
21. Tinnakornsrisuphap T, Billo RE (2015) An interoperable system for automated diagnosis of cardiac abnormalities from electrocardiogram data. IEEE J Biomed Health Inform 19(2):493–500
22. Bigdely-Shamlo N, Cockfield J, Makeig S, Rognon T, La Valle C, Miyakoshi M, Robbins KA (2016) Hierarchical Event Descriptors (HED): semi-structured tagging for real-world events in large-scale EEG. Front Neuroinform 10:42
23. Li H, Wu J, Gao Y, Shi Y (2016) Examining individuals' adoption of healthcare wearable devices: an empirical study from privacy calculus perspective. Int J Med Inform 88:8–17
24. McCarthy M (2016) Federal privacy rules offer scant protection for users of health apps and wearable devices. BMJ (Clinical Res Ed) 354:i4115
25. Safavi S, Shukur Z (2014) Conceptual privacy framework for health information on wearable device. PLoS One 9(12):e114306
26. Wu E, Torous J, Hardaway R, Gutheil T (2017) Confidentiality and privacy for smartphone applications in child and adolescent psychiatry: unmet needs and practical solutions. Child Adolesc Psychiatr Clin N Am 26(1):117–124
27. Zhu H, Liu X, Lu R, Li H (2017) Efficient and privacy-preserving online medical prediagnosis framework using nonlinear SVM. IEEE J Biomed Health Inform 21(3):838–850
28. Bayasi N, Tekeste T, Saleh H, Mohammad B, Khandoker A, Ismail M (2016) Low-power ECG-based processor for predicting ventricular arrhythmia. IEEE Trans Very Large Scale Integr (VLSI) Syst 24(5):1962–1974
29. Koelstra S, Muhl C, Soleymani M, Lee J-S, Yazdani A, Ebrahimi T, Pun T, Nijholt A, Patras I (2012) DEAP: a database for emotion analysis using physiological signals. IEEE Trans Affect Comput 3(1):18–31
30. Kim YG, Shin D, Park MY, Lee S, Jeon MS, Yoon D, Park RW (2017) ECG-ViEW II, a freely accessible electrocardiogram database. PLoS One 12(4):e0176222
31. Jenkins JM, Jenkins RE (2003) Arrhythmia database for algorithm testing: surface leads plus intracardiac leads for validation. J Electrocardiol 36:157–161
32. Mukkamala R, Moody GB, Mark RG (2001) Introduction of computational models to PhysioNet. Comput Cardiol 28:77–80
33. Goldberger AL, Amaral LA, Glass L, Hausdorff JM, Ivanov PC, Mark RG, Mietus JE, Moody GB, Peng CK, Stanley HE (2000) PhysioBank, PhysioToolkit, and PhysioNet: components of a new research resource for complex physiologic signals. Circulation 101(23):E215–E220
34. Moody GB, Mark RG, Goldberger AL (2000) PhysioNet: a research resource for studies of complex physiologic and biomedical signals. Comput Cardiol 27:179–182
35. Moody GB, Mark RG, Goldberger AL (2001) PhysioNet: a web-based resource for the study of physiologic signals. IEEE Eng Med Biol Mag 20(3):70–75
36. Costa M, Moody GB, Henry I, Goldberger AL (2003) PhysioNet: an NIH research resource for complex signals. J Electrocardiol 36(Suppl):139–144
37. Obeid I, Picone J (2016) The Temple University Hospital EEG data corpus. Front Neurosci 10:196
38. Devuyst S, Dutoit T, Stenuit P, Kerkhofs M (2011) Automatic sleep spindles detection – overview and development of a standard proposal assessment method. Conference proceedings: Annual International Conference of the IEEE Engineering in Medicine and Biology

Society IEEE Engineering in Medicine and Biology Society Annual Conference 2011:1713–1716
39. Tsanas A, Clifford GD (2015) Stage-independent, single lead EEG sleep spindle detection using the continuous wavelet transform and local weighted smoothing. Front Hum Neurosci 9:15
40. Khandelwal S, Wickstrom N (2017) Evaluation of the performance of accelerometer-based gait event detection algorithms in different real-world scenarios using the MAREA gait database. Gait Posture 51:84–90
41. Balazia M, Plataniotis KN (2017) Human gait recognition from motion capture data in signature poses. Iet Biom 6(2):129–137
42. Herrera LJ, Fernandes CM, Mora AM, Migotina D, Largo R, Guillen A, Rosa AC (2013) Combination of heterogeneous EEG feature extraction methods and stacked sequential learning for sleep stage classification. Int J Neural Syst 23(3):1350012
43. Romhilt DW, Estes EH Jr (1968) A point-score system for the ECG diagnosis of left ventricular hypertrophy. Am Heart J 75(6):752–758
44. Skjaeggestad O, Kierulf P (1971) A simplified point score system for the electrocardiographic diagnosis of left ventricular hypertrophy. Acta Med Scand 190(6):527–529
45. Sivaraks H, Ratanamahatana CA (2015) Robust and accurate anomaly detection in ECG artifacts using time series motif discovery. Comput Math Methods Med 2015:453214
46. Takahashi T, Cho RY, Mizuno T, Kikuchi M, Murata T, Takahashi K, Wada Y (2010) Antipsychotics reverse abnormal EEG complexity in drug-naive schizophrenia: a multiscale entropy analysis. NeuroImage 51(1):173–182
47. Abbasi H, Bennet L, Gunn AJ, Unsworth CP (2017) Robust wavelet stabilized 'Footprints of Uncertainty' for fuzzy system classifiers to automatically detect sharp waves in the EEG after hypoxia ischemia. Int J Neural Syst 27(3):1650051
48. Kim J, Hyub H, Yoon SZ, Choi HJ, Kim KM, Park SH (2014) Analysis of EEG to quantify depth of anesthesia using hidden Markov model. Conf Proc IEEE Eng Med Biol Soc 2014:4575–4578
49. Amin HU, Malik AS, Ahmad RF, Badruddin N, Kamel N, Hussain M, Chooi WT (2015) Feature extraction and classification for EEG signals using wavelet transform and machine learning techniques. Australas Phys Eng Sci Med 38(1):139–149
50. Mahadevan A, Mugler DH, Acharya S (2008) Adaptive filtering of ballistocardiogram artifact from EEG signals using the dilated discrete Hermite transform. Conf Proc IEEE Eng Med Biol Soc 2008:2630–2633
51. Oweis RJ, Abdulhay EW (2011) Seizure classification in EEG signals utilizing Hilbert-Huang transform. Biomed Eng Online 10:38
52. Poh KK, Marziliano P (2007) Analysis of neonatal EEG signals using Stockwell transform. Conf Proc IEEE Eng Med Biol Soc 2007:594–597
53. Thuraisingham RA, Tran Y, Craig A, Nguyen H (2012) Frequency analysis of eyes open and eyes closed EEG signals using the Hilbert-Huang transform. Conf Proc IEEE Eng Med Biol Soc 2012:2865–2868
54. Vidal F, Burle B, Spieser L, Carbonnell L, Meckler C, Casini L, Hasbroucq T (2015) Linking EEG signals, brain functions and mental operations: advantages of the Laplacian transformation. Int J Psychophysiol Off J Int Organ Psychophysiol 97(3):221–232
55. Chan NY, Choy CC (2017) Screening for atrial fibrillation in 13 122 Hong Kong citizens with smartphone electrocardiogram. Heart 103(1):24–31
56. Jain SK, Bhaumik B (2017) An energy efficient ECG signal processor detecting cardiovascular diseases on smartphone. IEEE Trans Biomed Circ Syst 11(2):314–323
57. Son J, Park J, Oh H, Bhuiyan MZA, Hur J, Kang K (2017) Privacy-preserving electrocardiogram monitoring for intelligent arrhythmia detection. Sensors (Basel) 17(6):E1360
58. Muhlestein JB, Le V, Albert D, Moreno FL, Anderson JL, Yanowitz F, Vranian RB, Barsness GW, Bethea CF, Severance HW et al (2015) Smartphone ECG for evaluation of STEMI: results of the ST LEUIS pilot study. J Electrocardiol 48(2):249–259

59. Yang H, Fayad A, Chaput A, Oake S, Chan AD, Crossan ML (2017) Postoperative real-time electrocardiography monitoring detects myocardial ischemia: a case report. Canadian J Anaesth=J Can Anesth 64(4):411–415
60. Melillo P, Castaldo R, Sannino G, Orrico A, de Pietro G, Pecchia L (2015) Wearable technology and ECG processing for fall risk assessment, prevention and detection. Conf Proc IEEE Eng Med Biol Soc 2015:7740–7743
61. Peritz DC, Howard A, Ciocca M, Chung EH (2015) Smartphone ECG aids real time diagnosis of palpitations in the competitive college athlete. J Electrocardiol 48(5):896–899
62. Karlen W, Mattiussi C, Floreano D (2009) Sleep and wake classification with ECG and respiratory effort signals. IEEE Trans Biomed Circuits Syst 3(2):71–78
63. Wang Y, Chen JJ, Li QH, Wang HY, Liu GQ, Jing Q, Shen BR (2011) Identifying novel prostate cancer associated pathways based on integrative microarray data analysis. Comput Biol Chem 35(3):151–158
64. Tang YF, Yan WY, Chen JJ, Luo C, Kaipia A, Shen BR (2013) Identification of novel microRNA regulatory pathways associated with heterogeneous prostate cancer. BMC Syst Biol 7:S6
65. Hu YF, Li JQ, Yan WY, Chen JJ, Li Y, Hu G, Shen BR (2013) Identifying novel glioma associated pathways based on systems biology level meta-analysis. BMC Syst Biol 7:S9
66. Chen JJ, Zhang DQ, Yan WY, Yang DR, Shen BR (2013) Translational bioinformatics for diagnostic and prognostic prediction of prostate cancer in the next-generation sequencing era. Biomed Res Int 2013:901578
67. Chen JJ, Sun MM, Shen BR (2015) Deciphering oncogenic drivers: from single genes to integrated pathways. Brief Bioinform 16(3):413–428
68. Chen JJ, Wang Y, Guo DY, Shen BR (2012) A systems biology perspective on rational design of peptide vaccine against virus infections. Curr Top Med Chem 12(12):1310–1319
69. Lin YX, Yuan XY, Shen BR (2016) Network-based biomedical data analysis. In: Shen B, Tang H, Jiang X (eds) Translational biomedical informatics: a precision medicine perspective, vol 939. Springer, Singapore, pp 309–332
70. Shen K, Shen L, Wang J, Jiang Z, Shen B (2015) Understanding amino acid mutations in hepatitis B virus proteins for rational design of vaccines and drugs. Adv Protein Chem Struct Biol 99:131–153
71. Moss AJ, Zareba W, Benhorin J, Locati EH, Hall WJ, Robinson JL, Schwartz PJ, Towbin JA, Vincent GM, Lehmann MH (1995) ECG T-wave patterns in genetically distinct forms of the hereditary long QT syndrome. Circulation 92(10):2929–2934
72. Vanninen SUM, Nikus K, Aalto-Setala K (2017) Electrocardiogram changes and atrial arrhythmias in individuals carrying sodium channel SCN5A D1275N mutation. Ann Med 49(6):496–503
73. Moeller F, Groening K, Moehring J, Muhle H, Wolff S, Jansen O, Stephani U, Siniatchkin M (2014) EEG-fMRI in myoclonic astatic epilepsy (doose syndrome). Neurology 82(17):1508–1513
74. Perlaki G, Orsi G, Schwarcz A, Bodi P, Plozer E, Biczo K, Aradi M, Doczi T, Komoly S, Hejjel L et al (2015) Pain-related autonomic response is modulated by the medial prefrontal cortex: an ECG-fMRI study in men. J Neurol Sci 349(1–2):202–208
75. Arns M, Gordon E, Boutros NN (2017) EEG abnormalities are associated with poorer depressive symptom outcomes with escitalopram and venlafaxine-XR, but not sertraline: results from the multicenter randomized iSPOT-D study. Clin EEG Neurosci 48(1):33–40
76. Lee SH, Yoon S, Kim JI, Jin SH, Chung CK (2014) Functional connectivity of resting state EEG and symptom severity in patients with post-traumatic stress disorder. Prog Neuro-Psychopharmacol Biol Psychiatry 51:51–57
77. Peng Q, Schork NJ, Wilhelmsen KC, Ehlers CL (2017) Whole genome sequence association and ancestry-informed polygenic profile of EEG alpha in a native American population. Am J Med Genet Part B, Neuropsychiatr Genet Off Publ Int Soc Psychiatr Genet 174(4):435–450

78. Bonanni L, Franciotti R, Nobili F, Kramberger MG, Taylor JP, Garcia-Ptacek S, Falasca NW, Fama F, Cromarty R, Onofrj M et al (2016) EEG markers of dementia with Lewy bodies: a multicenter cohort study. J Alzheim Dis JAD 54(4):1649–1657
79. Hautz WE, Krummrey G, Exadaktylos A, Hautz SC (2016) Six degrees of separation: the small world of medical education. Med Educ 50(12):1274–1279
80. Wicks P, Massagli M, Frost J, Brownstein C, Okun S, Vaughan T, Bradley R, Heywood J (2010) Sharing health data for better outcomes on PatientsLikeMe. J Med Internet Res 12(2): e19
81. Wang L, Fang Y, Aref D, Rathi S, Shen L, Jiang X, Wang S (2016) PALME: PAtients Like My gEnome. AMIA Joint Summits Transl Sci Proc AMIA Joint Summits Transl Sci 2016:219–224
82. Engel GL (2012) The need for a new medical model: a challenge for biomedicine. Psychodyn Psychiatry 40(3):377–396
83. Collins FS, Fleming R (2017) Sound health: an NIH-Kennedy center initiative to explore music and the mind. JAMA

# Chapter 3
# Entropy for the Complexity of Physiological Signal Dynamics

Xiaohua Douglas Zhang

**Abstract** Recently, the rapid development of large data storage technologies, mobile network technology, and portable medical devices makes it possible to measure, record, store, and track analysis of biological dynamics. Portable noninvasive medical devices are crucial to capture individual characteristics of biological dynamics. The wearable noninvasive medical devices and the analysis/management of related digital medical data will revolutionize the management and treatment of diseases, subsequently resulting in the establishment of a new healthcare system. One of the key features that can be extracted from the data obtained by wearable noninvasive medical device is the complexity of physiological signals, which can be represented by entropy of biological dynamics contained in the physiological signals measured by these continuous monitoring medical devices. Thus, in this chapter I present the major concepts of entropy that are commonly used to measure the complexity of biological dynamics. The concepts include Shannon entropy, Kolmogorov entropy, Renyi entropy, approximate entropy, sample entropy, and multiscale entropy. I also demonstrate an example of using entropy for the complexity of glucose dynamics.

**Keywords** High-throughput phenotyping • Entropy • Complexity • Wearable medical device • Continuous monitoring

## 3.1 Introduction

In this century, the rapid development of genomics biotechnologies, large data storage technologies, mobile network technology, and portable medical devices makes it possible to measure, record, store, and track analysis of the genome, physiological dynamics, and living environment and subsequently to reveal the differences and uniqueness of an individual. These technologies may allow us to reclassify the diseases for making diagnostic and therapeutic strategies more precisely tailored to individual patients, leading the birth of precision medicine and personalized medicine. Portable noninvasive medical devices are crucial to

X.D. Zhang (✉)
Faculty of Health Sciences, University of Macau, Taipa, Macau, China
e-mail: douglaszhang@umac.mo

capture individual characteristics of biological dynamics. In fact, the rapid development of wearable, mobile, automatic, continuous, high-throughput medical device for measuring human biological parameters heralds a new era [5] – high throughput phenotyping era. In this new era, the wearable noninvasive m5edical devices and the analysis/management of related digital medical data will revolutionize the management and treatment of diseases, subsequently resulting in the establishment of a new healthcare system including those for the treatment and care of respiratory patients.

One of the key features that can be extracted from the data obtained by the high-throughput medical device is the complexity of physiological signals. Thus, complexity studies in various major diseases including respiratory diseases are arising as an important method to analyze these continuous monitoring data measured by the noninvasive medical devices. The complexity of physiological signals can be represented by entropy of biological dynamics contained in the physiological signals measured by continuous monitoring medical devices.

The initial entropy is derived from and applied to the physics of thermodynamics and statistical physics. Clausius introduced the concept of entropy in the 1850s [1] and was the first one to enunciate the second law of thermodynamics by saying that "entropy always increases." Boltzmann was the first to state the logarithmic connection between entropy and probability in 1886. In 1948, Shannon [12] proposed an entropy (later known as Shannon entropy) and a large number of applications in information science. The Kolmogorov entropy [8] and Renyi entropy [10], which are developed on the basis of Shannon's entropy, are widely used in the nonlinear dynamics of the physical system. The entropy can be applied to the experimental data of biological dynamics, such as approximate entropy [9], sample entropy [11], and multiscale entropy [2, 14], to quantify the physiological signals in the physiological dynamic system, such as heart rate, airflow, pressure in airway, signal sound, and so on. Hence, in this chapter, I should describe the above major concepts of entropy, show their connections, and demonstrate an example of using entropy. The original concepts of entropy proposed by Clausius and Boltzmann are rarely used for biological dynamics directly. Thus, I start with Shannon entropy.

## 3.2 Shannon Entropy

For a discrete random variable $Y$ with a probability mass function $p(Y)$, the entropy is defined as the expectation of a function of $Y$, $I(Y)$, where $I(Y) = -\log(p(Y))$. That is,

$$H(Y) = E\{I(Y)\} = E\{-\log(p(Y))\} \qquad (3.1)$$

In information science, $I(Y)$ is the information content of $Y$, which is also a random variable. Suppose that the random variable $Y$ has possible values $\{y_1, y_2, \ldots, y_n\}$

and a corresponding probability mass function of $p_i = \Pr(Y = y_i)$. The entropy can be expanded as [12]

$$H(Y) = E\{I(Y)\} = E\{-\log(p(Y))\} = \sum_{i=1}^{n}(p_i \times (-\log(p_i)))$$

$$= -\sum_{i=1}^{n}(p_i \log p_i) \qquad (3.2)$$

This definition of entropy can be extended to a continuous random variable with a probability density function $f(Y)$ as follows:

$$H(Y) = E\{I(Y)\} = E\{-\log(f(Y))\} = -\int f(Y) \times \log(f(Y))dY \qquad (3.3)$$

The entropy defined on the discrete variable is most commonly used. Hence, we focus on the entropy defined on a discrete variable. The concept of entropy can be easily extended to multidimensional random variable.

Suppose there are two events, $X$ and $Y$, in question with $I$ possibilities for the first and $J$ for the second. Let $p(x_i, y_j)$ be the probability of the joint occurrence of $x_i$ for the first and $y_j$ for the second. The marginal mass density functions of $X$ and $Y$ are $p_X(x_i) = \sum_{j=1}^{J} p(x_i, y_j)$ and $p_Y(y_j) = \sum_{i=1}^{I} p(x_i, y_j)$, respectively. The entropy of the joint event is [12]

$$H(X,Y) = E_{X,Y}\{I(X,Y)\} = E_{X,Y}\{-\log(p(X,Y))\}$$

$$= -\sum_{i=1}^{I}\sum_{j=1}^{J}\left(p(x_i, y_j)\log\left(p(x_i, y_j)\right)\right) \qquad (3.4)$$

The entropy of $X$ and $Y$ are, respectively,

$$H(X) = E_X\{-\log(p(X))\} = -\sum_{i=1}^{I}(p_X(x_i)\log(p_X(x_i)))$$

$$= -\sum_{i=1}^{I}\sum_{j=1}^{J}\left(p(x_i, y_j)\log(p_X(x_i))\right) = -\sum_{i=1}^{I}\sum_{j=1}^{J}\left(p(x_i, y_j)\log\left(\sum_{j=1}^{J}p(x_i, y_j)\right)\right)$$

$$H(Y) = E_Y\{-\log(p(Y))\} = -\sum_{j=1}^{J}\left(p_Y(y_j)\log\left(p_Y(y_j)\right)\right)$$

$$= -\sum_{j=1}^{J}\sum_{i=1}^{I}\left(p(x_i, y_j)\log\left(p_Y(y_j)\right)\right) = -\sum_{i=1}^{I}\sum_{j=1}^{J}\left(p(x_i, y_j)\log\left(p_Y(y_j)\right)\right)$$

$$= -\sum_{i=1}^{I}\sum_{j=1}^{J}\left(p(x_i, y_j)\log\left(\sum_{i=1}^{I}p(x_i, y_j)\right)\right)$$

The conditional probability of $Y = y_j$ given $X = x_i$ is $p(Y|X) = \dfrac{p(x_i y_j)}{\sum_{j=1}^{n} p(x_i y_j)} = \dfrac{p(x_i y_j)}{p_X(x_i)}$

The conditional entropy of two discrete random variables $X$ and $Y$ is defined as [12]

$$H_X(Y) = \mathrm{E}_{X,Y}\{I(Y|X)\} = \mathrm{E}_{X,Y}\{-\log(p(Y|X))\}$$

$$= -\sum_{i=1}^{I}\sum_{j=1}^{J}\left(p(x_i, y_j)\log\dfrac{p(x_i, y_j)}{\sum_{j=1}^{n} p(x_i, y_j)}\right) \quad (3.5)$$

Based on Shannon [12], the entropy has the following major properties for serving as a measure of choice or information:

(i) $H = 0$ if and only if all the $p_i$ but one are zero, this one having the value unity. Thus only when we are certain of the outcome does H vanish. Otherwise H is positive.
(ii) For a given $n$, H is a maximum and equal to $\log n$ when all the $p_i$ are equal (i.e., $\frac{1}{n}$). This is also intuitively the most uncertain situation.
(iii) Any change toward equalization of the probabilities $p_1, p_2, \ldots, p_n$ increases H. Thus if $p_1 < p_2$ and we increase $p_1$, decreasing $p_2$ an equal amount so that $p_1$ and $p_2$ are more nearly equal, then H increases. More generally, if we perform any "averaging" operation on the $p_i$ of the form $p'_i = \sum_{j}^{n} a_{ij} p_j$ where $\sum_{j}^{n} a_{ij} = \sum_{i}^{n} a_{ij} = 1$ and all $a_{ij} \geq 0$, then H increases (except in the special case where this transformation amounts to no more than a permutation of the $p_i$ with H of course remaining the same).
(iv) It is easily shown that $H(X,Y) \leq H(X) + H(Y)$ because $p(x_i, y_j) \geq p_X(x_i)p_Y(y_j)$. The equality holds only if the events are independent (i.e., $p(x_i, y_j) = p_X(x_i) p_Y(y_j)$). The uncertainty of a joint event is less than or equal to the sum of the individual uncertainties.
(v) It is easily shown that $H(X,Y) = H(X) + H_X(Y)$. The uncertainty (or entropy) of the joint event $X$ and $Y$ is the uncertainty of $X$ plus the uncertainty of $Y$ given $X$ is known.
(vi) $H(Y) \geq H_X(Y)$ because $H(X) + H(Y) \geq H(X,Y) = H(X) + H_X(Y)$. The uncertainty of $Y$ is never increased by knowledge of $X$. It will be decreased unless $X$ and $Y$ are independent events, in which case it is not changed.

## 3.3 Conditional Entropy

From Shannon's definition [12] leading to the relationship of $H(X,Y) = H(X) + H(Y|X)$, i.e.,

$$H(X,Y) = -\sum_{i=1}^{m}\sum_{j=1}^{n}\left(p(x_i,y_j)\log\left(p(x_i,y_j)\right)\right)$$

$$= -\sum_{i=1}^{m}\sum_{j=1}^{n}\left(p(x_i,y_j)\log\left(\sum_{j=1}^{n}p(x_i,y_j)\times\frac{p(x_i,y_j)}{\sum_{j=1}^{n}p(x_i,y_j)}\right)\right)$$

$$= -\sum_{i=1}^{m}\sum_{j=1}^{n}\left(p(x_i,y_j)\log\left(\sum_{j=1}^{n}p(x_i,y_j)\right)\right)$$

$$-\sum_{i=1}^{m}\sum_{j=1}^{n}\left(p(x_i,y_j)\log\left(\frac{p(x_i,y_j)}{\sum_{j=1}^{n}p(x_i,y_j)}\right)\right) = H(X) + H_X(Y)$$

we know that $X$ is not fixed at $x_i$ when the conditional entropy $H(Y|X)$ is not for one fixed $x_i$.

It is interesting to see that, if $Z_i = Y|(X = x_i)$ is treated as a random variable, the entropy defined on $Z_i$ will be

$$H(Z_i) = H(Y|X = x_i) = E_{Z_i}\{I(Z_i)\} = E_{Z_i}\{-\log(p(Z_i))\}$$

$$= -\sum_{j=1}^{n}\left(p(y_j|x_i)\log\left(p(y_j|x_i)\right)\right)$$

$$= -\sum_{j=1}^{n}\left(\frac{p(x_i,y_j)}{\sum_{j=1}^{n}p(x_i,y_j)}\log\left(\frac{p(x_i,y_j)}{\sum_{j=1}^{n}p(x_i,y_j)}\right)\right)$$

which differs from the conditional entropy of two discrete random variable. Note, here $X$ is fixed at $x_i$. In another word, the conditional entropy differs from the entropy of a conditional probability or event. The conditional entropy of $Y$ given $X$ measures the uncertainty of $Y$ given we know $X$ regardless of what the value of $X$ is, whereas the entropy of $Y$ given $X$ measures the uncertainty of $Y$ given $X$ equals one of its specific values. Their relationship can be shown below:

$$H_X(Y) = -\sum_{i=1}^{m}\sum_{j=1}^{n}\left(p(x_i,y_j)\log\left(p(y_j|x_i)\right)\right)$$

$$= -\sum_{i=1}^{m}\sum_{j=1}^{n}\left(p(x_i)p(y_j|x_i)\log p(y_j|x_i)\right)$$

$$= -\sum_{i=1}^{m}\left(p(x_i)\sum_{j=1}^{n}p(y_j|x_i)\log p(y_j|x_i)\right) = -\sum_{i=1}^{m}(p(x_i)H(Z_i))$$

That is,

$$H_X(Y) = -\sum_{i=1}^{m}(p(x_i)H(Z_i)) \tag{3.6}$$

Hence, the conditional entropy is a weighted or average entropy of conditional probabilities or events.

## 3.4 Renyi Entropy

In mathematics, Shannon entropy can be seen as a special case of Renyi entropy [10] which is introduced here. In general, for a discrete random variable $Y$ with a probability mass function $p_i = \Pr(Y = y_i)$, the Renyi entropy of order $q$, where $q \geq 0, q \neq 1$, is

$$R_q(Y) = -\frac{1}{q-1}\log\sum_{i=1}^{n}p_i^q \tag{3.7}$$

The limit of Renyi entropy for $q \to 1$ is the Shannon entropy, that is,

$$\lim_{q \to 1} R_q(Y) = -\lim_{q \to 1}\frac{1}{q-1}\log\sum_{i=1}^{n}p_i^q = -\sum_{i=1}^{n}(p_i \log p_i) \tag{3.8}$$

Consequently, it can be defined $R_{q=1}(Y) = \lim_{q \to 1} R_q(Y)$. In such a case, the Shannon entropy is a special case of Renyi entropy with $q = 1$. It can be demonstrated that

$$R_{q_1}(Y) > R_{q_2}(Y) \text{ when } q_1 > q_2 \tag{3.9}$$

## 3.5 Kolmogorov Entropy

Kolmogorov entropy (also known as metric entropy) is originated from the field of dynamical systems, which is defined as follows [8]. Consider a dynamical system with a phase space. Divide the phase space into a set $N(c)$ of disjoint $D$-dimensional hypercubes of content $c^D$. Let $p(i_0, i_1, i_2, \ldots, i_N)$ be the probability that a trajectory is in hypercube $i_0$ at $t = 0$, hypercube $i_1$ at $t = \tau$, hypercube $i_2$ at $t = 2\tau \ldots$, hypercube $i_N$ at $t = N\tau$.

Kolmogorov entropy is then defined as

$$K = - \lim_{\tau \to 0} \lim_{\epsilon \to 0^+} \lim_{N \to \infty} \frac{1}{N\tau} \left( \sum_{i_0, i_1, i_2, \ldots, i_N} (p(i_0, i_1, i_2, \ldots, i_N) \log p(i_0, i_1, i_2, \ldots, i_N)) \right) \quad (3.10)$$

The above definition relies on quite some jargons such as phase space, hypercube, content, trajectory, and so on in physics that a statistician is usually unfamiliar.

A dynamical system is a description of a physical system that evolves over time. The system has many states, and all states are represented in the state space of the system. The state space is also termed as phase space. A path in the state space describes the dynamics of the dynamical system. The path is termed as trajectory.

Kolmogorov entropy is the theoretical basis for approximate entropy, sample entropy, and multiscale entropy that are commonly used in time series, which relies on the concept that Kolmogorov entropy is the rate of change of Shannon entropy of a system. The Kolmogorov-Sinai entropy measures unpredictability of a dynamical system. The higher the unpredictability, the higher the entropy. A higher Kolmogorov entropy value means a higher rate of change of the internal structure and of the information content and thus the faster development of complexity. The following paragraph should help us to understand why Kolmogorov entropy is the rate of change of Shannon entropy of a system.

Based on the setting for Eq. 3.13, Shannon entropy for a fix $n$ is

$$K_n = - \sum_{i_0, i_1, i_2, \ldots, i_n} (p(i_0, i_1, i_2, \ldots, i_n) \log p(i_0, i_1, i_2, \ldots, i_n)) \quad (3.11)$$

Then, $K_{n+1} - K_n$ is the increment of Shannon entropy from time $n\tau$ to $(n+1)\tau$, which can be seen as the information needed to predict the status at $(n+1)\tau$ given the status at up to $n\tau$ is known. The overall rate of change $H'$ of Shannon entropy is thus

$$H' = \lim_{\tau \to 0} \lim_{\epsilon \to 0^+} \lim_{N \to \infty} \frac{1}{N\tau} \sum_{n=0}^{N-1} (K_{n+1} - K_n) \quad (3.12)$$

Consider

$$\sum_{n=0}^{N-1} (K_{n+1} - K_n) = (K_1 - K_0) + (K_2 - K_1) + (K_3 - K_2) + \ldots + (K_N - K_{N-1})$$

$$= K_N - K_0$$

and $K_0 = \sum_{i_0} (p(i_0) \log p(i_0)) = 0$ because status at $t = 0$ is always known and $p(i_0) = 1$, we have

$$H' = \lim_{\tau \to 0} \lim_{\epsilon \to 0^+} \lim_{N \to \infty} \frac{1}{N\tau} \sum_{n=0}^{N-1} (K_{n+1} - K_n) = \lim_{\tau \to 0} \lim_{\epsilon \to 0^+} \lim_{N \to \infty} \frac{1}{N\tau}(K_N - K_0) = \lim_{\tau \to 0} \lim_{\epsilon \to 0^+} \lim_{N \to \infty} \frac{1}{N\tau} K_N$$

$$= -\lim_{\tau \to 0} \lim_{\epsilon \to 0^+} \lim_{N \to \infty} \frac{1}{N\tau} \left( \sum_{i_0, i_1, i_2, \ldots, i_N} (p(i_0, i_1, i_2, \ldots, i_N) \log p(i_0, i_1, i_2, \ldots, i_N)) \right) = K$$

That is, $K = H'$, which shows that Kolmogorov entropy is the rate of change of Shannon entropy of a system.

The application of Kolmogorov entropy in time series commonly goes through another quantity $R_2$, which is a special case of the rate of change of Renyi entropy $R_q$.

For the setting for Eq. 3.9, the rate of change of Renyi entropy is

$$R_q = -\lim_{\tau \to 0} \lim_{\epsilon \to 0} \lim_{N \to \infty} \frac{1}{N\tau} \frac{1}{q-1} \log \sum_{i_0, i_1, i_2, \ldots, i_N} p^q(i_0, i_1, i_2, \ldots, i_N) \qquad (3.13)$$

From Eqs. 3.7 and 3.8, we have that $R_1$ is the Kolmogorov entropy and $R_2$ is the lower bound of Kolmogorov entropy.

Grassberger and Procaccia [6] demonstrated that for typical cases, $R_2$ is numerically close to $K$. More importantly, $R_2$ can be extracted fairly easily from an experimental signal as follows.

Let $X_i = (y_i, y_{i+1}, \ldots, y_{i+d-1})$ (where $i$ is from $1$ to $N-d+1$) be a sequence of $Y$ starting at $y_i$ with a length of $d$. That is, we have a sequence of $N-d+1$ vectors, $X_1$, $X_2, \ldots, X_{N-d+1}$.

Consider

$$C_d(\epsilon) = \lim_{N \to \infty} \left\{ \frac{1}{N^2} \times \text{number of pairs}(n, m) \text{with} \left( \sum_{i=1}^{d} (X_{n+i} - X_{m+i})^2 \right)^{1/2} < \epsilon \right\}$$

$$K_{2,d}(\epsilon) = \frac{1}{\tau} \log \frac{C_d(\epsilon)}{C_{d+1}(\epsilon)}$$

Grassberger and Procaccia [6] proved that

$$\lim_{\substack{d \to \infty \\ \epsilon \to 0}} K_{2,d}(\epsilon) = K_2, \text{ i.e.,} \quad \lim_{\substack{d \to \infty \\ \epsilon \to 0}} \frac{1}{\tau} \log \frac{C_d(\epsilon)}{C_{d+1}(\epsilon)} = K_2 \qquad (3.14)$$

The Euclidean distance in $C_d(\epsilon)$ may be replaced by the maximum norm [13].

## 3.6 Approximate Entropy

Consider a time series of data with length of $N$, $Y = (y_1, y_2, \ldots, y_N)$, from measurements equally spaced in time. Let $X_i = (y_i, y_{i+1}, \ldots, y_{i+m-1})$ (where $i$ is from $1$ to $N-m+1$) be a sequence of $Y$ starting at $y_i$ with a length of $m$. That is, we have a sequence of $N-m+1$ vectors, $X_1, X_2, \ldots, X_{N-m+1}$. For each pair of sequences, $X_i$ and $X_j$, we may define a distance between them. One commonly used distance is the maximum absolute difference of their corresponding elements, namely,

$$d(X_i, X_j) = \max(|X_i - X_j|) = \max_{k=1,2,\ldots,m} |y_{i+k-1} - y_{j+k-1}|$$

For a sequence $X_i$, a sequence $X_j$ that has a distance from $X_i$ less than or equal to $r$ (i.e., $d(X_i, X_j) \leq r$) is defined as within the r of $X_i$ and count it as a match with $X_i$. We may count the number of $X_j$ that matches with $X_i$ with respect to (w.r.t.) $r$ and denote it as $C_i^m(r)$. The proportion of $X_j$ matching with $X_i$ w.r.t. $r$ is then

$$P_i^m(r) = C_i^m(r)/(N - m + 1)$$

When $N$ is large, $P_i^m(r)$ represents the probability that any $X_j$ matching with $X_i$ w.r.t. $r$.

The average proportion of matches for all sequences $X_i$ ($1 \leq i \leq N - m + 1$) is thus

$$P^m(r) = \frac{1}{N - m + 1} \sum_{i=1}^{N-m+1} P_i^m(r)$$

Define $\Phi^m(r)$ as

$$\Phi^m(r) = \frac{1}{N - m + 1} \sum_{i=1}^{N-m+1} \log P_i^m(r)$$

Then, Eckmann-Ruelle (ER) entropy [4] is

$$\begin{aligned}
\text{ER entropy} &= \lim_{r \to 0} \lim_{m \to \infty} \lim_{N \to \infty} \left( \Phi^m(r) - \Phi^{m+1}(r) \right) \\
&= \lim_{r \to 0} \lim_{m \to \infty} \lim_{N \to \infty} \left( \frac{1}{N - m + 1} \sum_{i=1}^{N-m+1} \log P_i^m(r) - \frac{1}{N - m} \sum_{i=1}^{N-m} \log P_i^{m+1}(r) \right)
\end{aligned}$$

The approximate entropy for fixed $m$ and $r$ [9] is

$$\text{ApEn}(m,r) = \lim_{N\to\infty} \left(\Phi^m(r) - \Phi^{m+1}(r)\right)$$

$$= \lim_{N\to\infty} \left(\frac{1}{N-m+1}\sum_{i=1}^{N-m+1} \log P_i^m(r) - \frac{1}{N-m}\sum_{i=1}^{N-m} \log P_i^{m+1}(r)\right)$$

Given $N$ data points

$$\text{ApEn}(m,r,N) = \Phi^m(r) - \Phi^{m+1}(r)$$

$$= \frac{1}{N-m+1}\sum_{i=1}^{N-m+1} \log P_i^m(r) - \frac{1}{N-m}\sum_{i=1}^{N-m} \log P_i^{m+1}(r)$$

The theoretical framework for approximate entropy is based on the following theorems [9].

Assume a stationary process $u(i)$ with continuous state space. Let $(X, Y)$ be the joint stationary probability measure on a two-dimensional space for this process (assuming uniqueness), and $\pi_X$ be the equilibrium probability of X. Then a.s.

**Theorem 1**

$$\text{ApEn}(1,r) = -\int u(x,y)\log\left(\int_{z=y-r}^{y+r}\int_{w=x-r}^{x+r} u(w,z)\,dw\,dz \Big/ \int_{w=x-r}^{x+r} \pi(w)\,dw\right) dx\,dy$$

**Theorem 2** For an *i.i.d.* process with density function, a.s. (for any $m \geq 1$)

$$\text{ApEn}(m,r) = -\int \pi(y)\log\left(\int_{z=y-r}^{y+r} \pi(z)\,dz\right) dy$$

**Theorem 3** In the first-order stationary Markov chain (discrete state space values) case, with $r < \min(|x-y|, x \neq y, x$ and $y$ state space values $X$), a.s. for any $m$

$$\text{ApEn}(m,r) = -\sum_{x\in X}\sum_{y\in Y} \pi(x)p_{xy}\log p_{xy}$$

Pincus [9] considers ApEn($m, r$) as a family of formulas and ApEn($m, r, N$) as a family of statistics; system comparisons are intended with fixed $m$ and $r$. This family of statistics is rooted in the work of Grassberger and Procaccia (1983) [6, 7] and Eckmann and Ruelle [4]. The above theory and method for a measure of regularity proposed by Pincus [9] are closely related to the Kolmogorov entropy, the rate of generation of new information, which can be applied to the typically short and noisy time series of clinical data [11].

## 3.7 Sample Entropy

Like approximate entropy, sample entropy is also defined on the count $C_i^m(r)$ and proportion $P_i^m(r)$ of a sequence $X_j$ matching with another sequence $X_i$ w.r.t. $r$ but with two alterations [11]. First, self-matches are counted in the calculation of $C_i^m(r)$ for approximate entropy, whereas self-matches are excluded in the calculation of $C_i^m(r)$ for sample entropy. Second, for a fixed $m$, only the first $N-m$ sequences $X_j$ of length $m$ are used for calculating $C_i^m(r)$ to ensure that the number of sequences used for $C_i^m(r)$ is the same as the number of sequences available for $C_i^{m+1}(r)$. To avoid confusion, in the altered setting as contrasted to the unaltered sitting, we may use $C_i^m(r)^*$ to denote the number of matches with the $i^{th}$ sequence $X_i$ of length $m$, i.e., $C_i^m(r)^* = \{C_i^m(r); i = 1, 2, i-1, i+1, N-m\}$, and to use $C_i^m(r)^{**}$ to denote the number of matches with the $i^{th}$ sequence $X_i$ sequence of length $m+1$, i.e., $C_i^m(r)^{**} = \{C_i^{m+1}(r); i = 1, 2, i-1, i+1, N-m\}$. Similar to the proportion $P_i^m(r)$ of matches for approximate entropy, we have the proportions, $P_i^m(r)^*$ and $P_i^m(r)^{**}$, of matches for sample entropy as follows:

$$P_i^m(r)^* = C_i^m(r)^*/(N-m)$$

$$P_i^m(r)^{**} = C_i^m(r)^{**}/(N-m-1)$$

Similar to the average proportion $P^m(r)$ of matches for all sequences $X_i$ for approximate entropy, we have the average proportions, $P^m(r)^*$ and $P^m(r)^{**}$, of matches for sample entropy as follows:

$$P^m(r)^* = \frac{1}{N-m} \sum_{i=1}^{N-m} P_i^m(r)^*$$

$$P^m(r)^{**} = \frac{1}{N-m-1} \sum_{i=1}^{N-m-1} P_i^m(r)^{**}$$

Clearly, $P^m(r)^*$ is an estimate for the probability that two different sequences will match for $m$ points, whereas $P^m(r)^{**}$ is an estimate for the probability that two different sequences will match for $m+1$ points.

The sample entropy for fixed $m$ and $r$ [11] is defined as

$$\text{SampEn}(m,r) = \lim_{N \to \infty} \left( -\log \frac{P^m(r)^{**}}{P^m(r)^*} \right)$$

Given $N$ data points

$$\text{SampEn}(m, r, N) = -\log \frac{P^m(r)^{**}}{P^m(r)^*}$$

From the above definition, sample entropy is essentially the negative logarithm of the ratio of the probability that any two different sequences in a time series match for $m$ points to the probability that any two different sequences in the time series match for $m+1$ points.

The theoretical framework for sample entropy is based on Grassberger and Procaccia's (1983) work as concluded in Eq. 3.14.

## 3.8 Multiscale Entropy Analysis

Kolmogorov entropy, approximate entropy, and sample entropy described above are based on a single scale, reflecting the uncertainty of the next point given the past history of the series as their calculation depends on a function's one step difference (e.g., $K_{n+1}-K_n$). Hence, they do not account for features related to structure on scales other than the shortest one [2]. To address this issue, Zhang [14] introduced multiscale entropy to the analysis of the physical time series. Costa et al. [2] further extended the multiscale entropy technique applicable to the analysis of the biological time series.

The basic idea for calculating multiscale entropy is to first generate a new time series consisting of the means of consecutive nonoverlap segments each with fixed length $k$ of observed data points and then to calculate Kolmogorov entropy, approximate entropy, or sample entropy based on the newly generated time series. Specifically, we first divide the original signal ($\{y_i\}$, $1 \leq i \leq N$) into nonoverlapping segments of equal length ($k$) and calculating the mean value of the data points in each of these segments [14]. This process is called coarse graining, and the newly generated time series is called coarse-grained time series. The length $k$ is called a scale factor (or simply a scale). The coarse-graining process is repeated for a range of the scale factor. As the scale factor $k$ changes, we will construct different coarse-grained time series, and subsequently we will calculate corresponding entropy values on the newly coarse-grained time series. We may now plot the entropy of coarse-grained time series against the scale factor $k$. This procedure is called multiscale analysis (MSE) [2].

The coarse-graining process can be applied to not only the mean of a divided segment but also its variance and any moments in statistics [3]. Costa and Goldberger [3] termed the MSE on mean-based coarse-grained time series as $\text{MSE}_\mu$, MSE on variance-based coarse-grained time series as $\text{MSE}_\sigma^2$, and MSE on any moment-based coarse-grained time series as $\text{MSE}_n$. Note, in statistics, the mean is the first central moment, and the variance is the second central moment.

## 3.9 Demonstration of an Example

Diabetes is a major disease in human being. It is a group of metabolic diseases in which the patients' blood sugar level is high over a prolonged period. The key index for monitoring diabetes is blood glucose level in a person. The left panel of Fig. 3.1 shows the glucose levels of a healthy person (black points) and a patient with type II diabetes (gray points), measured once every 3 min in 2 days. I calculate the multiscale sample entropy based on the data shown in the left panel of Fig. 3.1 and obtain the results shown in the right panel of Fig. 3.1.

The average value and standard deviation of the glucose levels are 5.16 and 1.00, respectively, in the healthy person and insulin and 12.45 and 3.08, respectively, in the patient with type II diabetes. The diabetes patient has sample entropy values lower than the healthy person at each scale (the right panel of Fig. 3.1). The change of direction in entropy is different from that in the average value (or variability), which may indicate that entropy contains information that the average value cannot be disclosed.

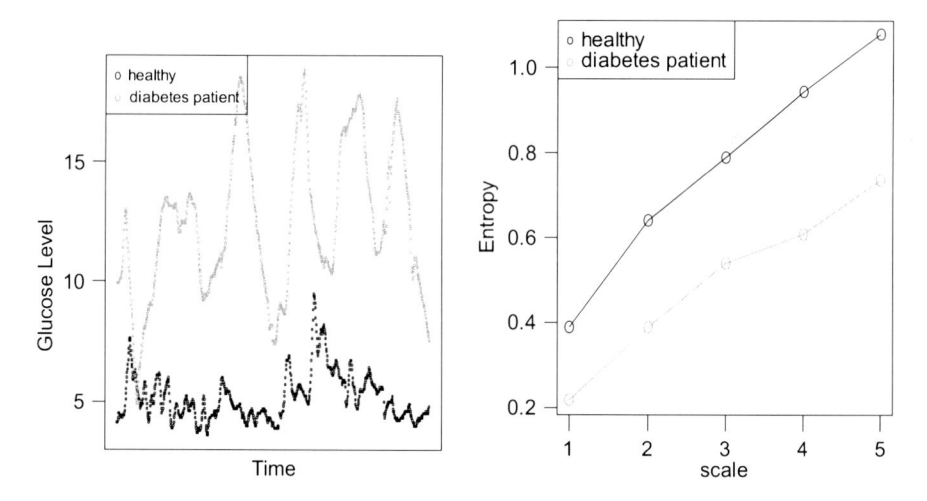

**Fig. 3.1** Human glucose dynamics and its analysis using multiscale sample entropy. *Left Panel*, value of glucose levels in a healthy person (*black points*) and in a patient with type II diabetes (*gray points*) measured once in every 3 min in 2 days. *Right Panel*, multiscale sample entropy values in a healthy person (*black points* and *line*) and in a patient with type II diabetes (*gray points*) calculated based on the glucose levels shown in the *left panel*

## 3.10 Discussion and Conclusion

Based on the description of various concepts of entropy for biological dynamics, the Shannon entropy is a measurement of the disorder (or uncertainty) of a system. Shannon entropy is a special case (with $q = 1$) of more broad entropy called Renyi entropy. Kolmogorov entropy is the rate of change of Shannon entropy. The low bound of Kolmogorov entropy is the rate of change of Renyi entropy with $q = 2$. This low bound can be approximated by an entropy that can be calculated using time series data from biological dynamics. This entropy is called approximate entropy. Sample entropy is an improved version of approximate entropy that corrects unpleased self-matching effect. Both sample entropy and approximate entropy can be calculated in different layers of data, resulting in multiscale entropy (MSE). Sample entropy, approximate entropy, and/or their corresponding MSE have now been broadly used to assess the complexity of biological dynamics in various diseases measured by noninvasive medical device.

As demonstrated in the diabetes example, entropy analysis for the dynamics of physiological signals can disclose information that is not contained in the average value or variability of the physiological signals. Thus, entropy analysis could be used to differentiate the major diseases such as diabetes into different sub-diseases on the top of existing approach (such as using the average value in the diabetes case). It could also be used to reclassify the major diseases so that more specific and more effective drugs or treatment can be developed. All these will help to the development of precision medicine in which the right drug at the right dosage can be prescribed to treat the right person at the right time.

## References

1. Clausius R (1854) Poggendoff's Ann., December 1854 xciii:481
2. Costa M, Goldberger AL, Peng CK (2002) Multiscale entropy analysis of complex physiologic time series. Phys Rev Lett 89(068):102
3. Costa M, Goldberger AL (2015) Generalized multiscale entropy analysis: application to quantifying the complex volatility of human heartbeat time series. Entropy 17:1197–1203
4. Eckmann JP, Ruelle D (1985) Ergodic theory of chaos and strange attractors. Rev Mod Phys 57:617–654
5. Elenko E, Underwood L, Zohar D (2015) Defining digital medicine. Nat Biotechnol 33:456–461
6. Grassberger P, Procaccia I (1983) Estimation of the Kolmogorov entropy from a chaotic signal. Phys Rev A 28:2591–2593
7. Grassberger P, Procaccia I (1983) Measuring the strangeness of strange attractors. Physica D 9:189–208
8. Kolmogorov AN (1958) New metric invariant of transitive dynamical systems and endomorphisms of Lebesgue spaces. Dokl Russ Acad Sci 119(N5):861–864
9. Pincus SM (1991) Approximate entropy as a measure of system complexity. Proc Natl Acad Sci U S A 88:2297–2301

10. Rényi A (1961) On measures of information and entropy. Proceedings of the fourth Berkeley Symposium on Mathematics, Statistics and Probability 1960. pp 547–561
11. Richman JS, Moorman JR (2000) Physiological time-series analysis using approximate entropy and sample entropy. Am J Physiol Heart Circ Physiol 278:H2039–H2049
12. Shannon CE (1948) A mathematical theory of communication. Bell Syst Tech J 27(3):379–423
13. Takens F (1983) Invariants related to dimension and entropy. Proceedings of the Thirteenth Coloquio Brasileiro de Matematicas (Rio de Janerio)
14. Zhang YC (1991) Complexity and 1/f noise. A phase space approach. J Phys I France 1:971–977

# Chapter 4
# Data Platform for the Research and Prevention of Alzheimer's Disease

Ning An, Liuqi Jin, Jiaoyun Yang, Yue Yin, Siyuan Jiang, Bo Jing, and Rhoda Au

**Abstract** With the rapid increase in global aging, Alzheimer's disease has become a major burden in both social and economic costs. Substantial resources have been devoted to researching this disease, and rich multimodal data resources have been generated. In this chapter, we discuss an ongoing effort to build a data platform to harness these data to help research and prevention of Alzheimer's disease. We will detail this data platform in terms of its architecture, its data integration strategy, and its data services. Then, we will consider how to leverage this data platform to accelerate risk factor identification and pathogenesis study with its data analytics capability. This chapter will provide a concrete pathway for developing a data platform for studying and preventing insidious onset chronic diseases in this data era.

**Keywords** Data platform • Alzheimer's disease • Implementation

## 4.1 Background

In this chapter, we focus on the data platform that could be used to advance the research and prevention of Alzheimer's disease. At the beginning, some background about Alzheimer's disease will be introduced, including its social affects, biomarkers, epidemiological investigation, etc.

### 4.1.1 Alzheimer's Disease

Alzheimer's disease (AD) is a chronic neurodegenerative disease that usually starts slowly and gets worse over time, and it causes about 60–70% of cases of dementia.

---

N. An • L. Jin • J. Yang (✉) • Y. Yin • S. Jiang • B. Jing
School of Computer and Information, Hefei University of Technology, 193 Tunxi Road, Hefei, Anhui 230009, China
e-mail: jiaoyun@hfut.edu.cn

R. Au
School of Medicine, Boston University, 72 E. Concord Street, Boston, MA 02118, USA

As medical science continues to reduce mortality, there is a concomitant increase in the risk for Alzheimer's disease (AD) and other types of dementia. A projected 13.8 million people in the United States are estimated to be diagnosed with AD by the year 2050 [1]. Despite this staggering number, the larger, more stunning global impact of this disease may well come from Third World countries where the health advances coupled with population growth strategies are converging to create a disproportionately large elderly population. In the year 2015, there were about 46.8 million people with dementia worldwide, and estimates are for 131.5 million by 2050, the majority of which will be diagnosed with Alzheimer's disease [37].

Besides, as a social disease, AD brings great burden to society and patients' family. From 2000 to 2010, the cost of caring for those with dementia/AD rose from $18 billion to over $600 billion. The burden of caring for people with AD is estimated to be $1.2 trillion by 2050 [2, 3]. It has become a serious disease affecting the global public health and attracted great attentions from global researchers. Numerous works have been done on this, for example, there are more than 112,000 research papers in PubMed up to 2015. Unfortunately, to date, there are still no disease-modifying medications. In order to tackle AD, the Obama administration and the National Institutes of Health have homed in on Alzheimer's disease, setting an ambitious goal to have an effective treatment for the brain-wasting disease by 2025.

Actually, some remarkable findings have been revealed, such as some biomarkers to determine AD's pathogenesis. Besides, research indicates that delaying onset by just 5 years will cut an individual's risk for diagnosis in half [4]. Identifying dementia risk factors that are amenable to intervention prior to disease onset represents a viable strategy to mitigate late-life dementia. And a number of dementia risk factor profiles have been developed from epidemiological studies. In the next two sections, we will introduce the biomarker findings and epidemiological investigations for AD.

### 4.1.2  Biomarkers for Alzheimer's Disease

A fundamental task for AD's research is to reveal the pathogenesis. Although the pathogenesis is still not clear, some biomarkers have been identified to help diagnose the early onset of the disease. This could guide the family to take some procedures to delay AD's deteriorating and help to reduce the cost of patients' healthcare.

Biomarkers could be extracted from various materials, such as blood, cerebrospinal fluid, MRI, PETs, genomes, etc. Current biomarkers that have been widely recognized include $A\beta$, Tau, APOE, etc.

- Abnormal amyloid-$\beta$-protein($A\beta$) aggregation
  Researchers injected $A\beta$ to the brains of transgenic mice, and found necrosis in the injected sites, missing in the surrounding neurons and glial proliferation [34]. Besides, these findings are closely related to the injected dose. After

culturing human brain cortex neurons and Aβ together, some Aβ formed amyloid-β-protein aggregation which would lead to neuronal degeneration. It's quite similar between neuronal degeneration by Aβ and pathologic changing of Alzheimer's disease.
- Hyper-phosphorylated tau protein aggregation

    The total amount of tau protein in the normal is less than that in AD's brains, whose normal tau protein decrease and hyper-phosphorylated tau protein increase largely. Hyper-phosphorylated tau in AD's brains loses not only functions of promoting microtubule assembly formation but also the ability of the adhesion strength of the microtubule protein, decreased by 90%. Compared with abnormal amyloid-β-protein (Aβ) aggregation, recent studies have shown that neurofibrillary tangles (NFTs) caused by hyper-phosphorylated tau protein have a higher correlation with AD [34].
- Apolipoprotein E (APOE) ε4

    Early in 1994, several studies have proved APOE ε4 has a meaningful relationship with praecox AD, and familial AD patients had a higher gene frequency of APOE ε4 than the nonfamilial patients. Comprehensive studies reveal the correlation of APOE ε4 and AD, which is that age of onset would bring forward because of the high amount of APOE ε4. For example, people who were homozygotes for APOE ε4 would have an earlier age of onset than people who were heterozygote for APOE ε4. Naturally, people with no APOE ε4 would have a later age of onset. Therefore, it can be recognized that APOE ε4 was one of the most important determinants of AD [35].

### *4.1.3 Epidemiological Investigation for Alzheimer's Disease*

While biomarkers' research aims to determine the pathogenesis and guide the treatment direction, prevention through modifiable risk factors remains the most cost-effective strategy for combatting the disease as evidenced by declining AD risk despite the lack of effective treatment options. Hence, various epidemiological investigations have been done to find AD-related risk factors.

Lifestyle risk factors including physical activity, sleep, and diet also impact risk for cognitive decline and AD. To a more limited extent, environmental exposures have also been linked to brain health. An important category of risk factors is cardiovascular risk factors (CVRF). Numerous studies suggest CVRF are associated with increased risk for cognitive decline and AD. Zlokovic's "two-hit vascular hypothesis of AD" states that vascular risk leads to neuronal injury and dysfunction (hit one), suggesting an initial non-amyloid pathway [9]. Identifying the specific role of potentially modifiable CVRF have on long-term cognitive health could lead to interventions that will allow lengthening years of active life and life quality. Delaying onset of CVRF by just 5 years cuts a person's risk by 50%.

Risk factors' discovery relies on epidemiological investigations by developing various protocols. Framingham Heart Study (FHS) is a pioneer in this area [5]. It was first initiated in 1948, and embarked on a study of heart disease, and is credited with identifying many of the known risk factors for cardiovascular disease. FHS has since applied its well-developed longitudinal study design to the investigation of prevalent and incident dementia and associated risk factors. The 2015 Alzheimer's Association estimates of lifetime risk of dementia and Alzheimer's disease utilized data from FHS' dementia study [6]. Some successful protocols they applied include socioeconomic, medical history questionnaire, blood and urine, physical activity, sleep questionnaire, food frequency questionnaire, environmental exposure, cardiovascular assessment, depression assessment, cognitive assessment, etc.

Traditional strategy for epidemiological investigation is usually based on questionnaire, and the investigation is usually done by face to face. However, more recent integration of wearable technology has allowed much more extensive assessment of physical activity and heart rate linked to cognitive impairment and brain structure, such as the Shimmer sensing platform could provide more complete, more continuous, and more accessible dynamics of the physiological signals associated with various levels and types of physical activity [7]. Besides, some digital devices could provide more accuracy assessment, e.g., the digital pen in place of a regular ballpoint pen for participant drawn tests [8]. The digital pen allows collection of decision-making latencies and graphomotor characteristics that may reflect subtle differences in underlying cognitive processing.

## 4.2 Public Data Platform

The biomarkers' research and epidemiological investigation have generated a large amount of data. These data comes from various institutes, and usually each institute could only make use of its own data. This limited data access may restrain the advance of this area on some levels, since some underlying knowledge could only be achieved when the data size reach a finite number. Therefore, some organizations turn to construct a uniform data platform to integrate these data from various institutes such that these numerous amounts of data could be fully utilized.

### 4.2.1 Public Data Platform for Alzheimer's Disease

- NACC (https://www.alz.washington.edu/)
   NACC is a public data platform established by the National Alzheimer's Coordinating Center. It provides services for the investigators of National Alzheimer's Centers, the officers of National Institute on Aging, staff of National Alzheimer's Coordinating Center, as well as the public [10].
   *Data contained*: NACC consists of three data sets, including Minimum Data Set (MDS), Neuropathology Data Set (NP), and Uniform Data Set (UDS). These

data sets cover data from questionnaires, medical images, genetics, and clinical examinations.

*Service provided*: data quality control, data share, data report generation, and data directory

- ADNI (http://adni.loni.usc.edu/)

  ADNI is built by Alzheimer's Disease Neuroimaging Initiative. Investigators could use the data to define each phrase of AD and to make prediction to AD. The data comes from ANDI's study participants, including Alzheimer's disease patients, mild cognitive impairment subjects, and the elderly [11, 12].

  *Data contained*: ADNI contains four types of data, including (1) clinical data, such as the recruitment information of each subject, demographic information, physical information, recognition assessment, CSF concentration, and the radiative data of biomarker; (2) MRI and PET image data, which could be used to track both the progression of AD and changes in the underlying pathology; (3) genetic data, such as genotyping and sequencing data; and (4) biospecimen data, such as blood, urine, and cerebrospinal fluid (CSF) data from participants.

  *Service provided*: data share, expert support, and data usage instruction

- AMP-AD Knowledge Portal (https://www.synapse.org)

  The platform is the knowledge portal for accelerating medicine partnership, AD target discovery, and preclinical validation. The main aim is to shorten the time between new medicine development and AD prevention. The platform provides a community for various consortiums to share data, data analysis method, and computing model [13].

  *Data contained*: RNA data, DNA data, genome data, protein data, and MRI data, among these data, some from human samples and some from animal samples

  *Service provided*: data usage agreement, data share, data analysis results share, data directory, besides, the platform provide query language to query the data; R, python, or command line could be applied to download the data file.

- The AddNeuroMed Study

  AddNeuroMed is funded by the EU FP6 program and is a public-private partnership for AD's biomarker discovery and replication. The platform aims to help the organization to share the data and integrate resources [14, 15].

  *Data contained*: (1) clinical data set, 700 samples; (2) DNA data set, collected from blood, 644 samples; (3) brain image data, 175 samples; (4) protein data, collected from blood, 645 samples; and (5) mRNA data, collected from blood, 674 samples

  *Service provided*: data usage agreement, data share, data query, and data analysis results share

- NIAGADS (https://www.niagads.org)

  National Institute on Aging Genetics of Alzheimer's Disease Data Storage Site (NIAGADS) is developed by NIA as a genetic data repository, which aims to share data for investigators studying the genetics of late-onset AD. The data is mainly used for investigators to do secondary research after the publication of any patent or research [16].

*Data contained*: genetic data generated from the analysis of patients' cerebrospinal fluid (patients' CHR, SNP), clinical examination data, and investigation data

*Service provided*: data share, data query, data comparison and annotation, web tools for analyzing large-scale genome data, and various software interfaces

- The Johns Hopkins Alzheimer's Disease Research Center

  ADRC is funded by NIA and established in 1984. The organization has made great contribution to the symptom and pathology research of AD. It aims to find AD's treatment in pathology. They provide both mental and physical ways to cure the patients [17].

  *Data contained*: blood and DNA data, brain specimen data, clinical data, and cognitive assessment data

  *Service provided*: data share and introduction to clinical research

- Alzheimer's Disease Cooperative Study

  ADCS is a cooperative organization between NIA and the University of California. It is the major Alzheimer's disease clinical research institute of the federal government, which aims to treat Alzheimer's disease for its physical and cognitive symptom [18].

  *Data contained*: cognitive assessment data and biomarkers' data

  *Service provided*: data share, integrated analysis tools to measure cognitive ability, life function, physical information, and quality of life

### 4.2.2 Classical Health-Related Data Platform

- CHARLS (http://charls.ccer.edu.cn)

  China Health and Retirement Longitudinal Study (CHARLS) aims to analyze Chinese aging problem and promote the interdisciplinary research based on the collection of a set of micro-high-quality data from typical families or persons older than 45 in China. The CHARLS baseline research began in 2011. It covers 150 counties and 450 villages and contains about 17,000 people from 10,000 families [19].

  *Data contained*: CHARLS consists of 12 data sets, including demographic background; family information; family transfer; health status and functioning; healthcare and insurance; work, retirement, and pension; household income; individual income; housing characteristics; interviewer observation; weights; and PSU.

  *Service Provided*: data share and data research results share

- cBioPortal

  The platform was first developed by the Memorial Sloan Kettering Cancer Center and now developed and maintained by multiple institutions including MSK, the Dana Farber Cancer Institute, Princess Margaret Cancer Centre in Toronto, Children's Hospital of Philadelphia, The Hyve in the Netherlands, and

Bilkent University in Ankara, Turkey. The platform aims to provide the visualization, analysis, and download of large-scale cancer genetic data [20].

*Data contained*: DNA copy-number data, mRNA, microRNA expression data, non-synonymous mutations, protein-level and phosphoprotein-level data, DNA methylation data, and limited de-identified clinical data

*Service provided*: data query, data share, webAPI (help access the txt or uml format data via uri), develop kit of R and MATLAB, visualization tools, and visualization tutorial

- CommonMind Consortium Knowledge Portal

    CMS Knowledge Portal aims to share the data and data analysis generated by CommonMind organization in a transparent, reproductive way. The platform commits to share the data, analysis results, and the source codes [21].

    *Data contained*: (1) DNA data set from schizophrenia, bipolar disorder, and mood disorder patients, 621 samples in total; (2) clinical data set that contains clinical data and metadata, 621 samples in total; and (3) RNA sequence data set from schizophrenia, bipolar disorder, and mood disorder patients, 613 samples in total

    *Service provided*: data share, data query, data analysis result share, and introduction to data analysis method

- US CDC Healthy Brain Initiative

    The platform is based on the healthy aging project of CDC, with the purpose of promoting the health and life quality of elder people. The initiative aims to integrate the public health and aging service network to promote the prevention of elder people's disease and promote the health-related life quality [22, 23].

    *Data contained*: general health data, nutrition, physical function data, vaccine relevant data, alcohol and smoking data, mental health data, and cognitive health data

    *Service provided*: data share, data visualization, API for data access, health education for the aged, and disease prevention

- NCMI (http://www.ncmi.cn)

    The National Scientific Data Sharing Platform for Population and Health (NCMI) serves for technology innovation, government administration, and development of medical and health service and provides data share service for cultivation of innovative talents and development of public health industry [24].

    *Data contained*: fundamental medical data, clinical medicine data, public health data, traditional Chinese medicine data, pharmacy data, population and reproductive health data

    *Service provided*: data share, data query, disease prevention, standard document, and data analysis tools (sequence and structure to predict mRNA)

- The Health Indicators Warehouse (http://www.healthindicators.gov/)

    The Health Indicators Warehouse (HIW) aims to use the high-quality data to help people understand the health status and determinants of community and to facilitate the prioritization of interventions. It could meet the need of various population health initiatives and provide a single, user-friendly data source for national, state, and community [25].

*Data contained*: health indicators including demographic, food health, situation of chronic disease, etc. and 1291 indicators in total

*Service provided*: data directory, data share, WEBAPI, and web service such as RESTFUL, SOAP, and EDGE

- The Michael J Fox for Parkinson (https://www.michaeljfox.org/)

    The platform is founded by Michael J. Fox Foundation and is dedicated to find strategies to cure the Parkinson disease via funded studies and ensured development of improved therapies [26].

    *Data contained*: DNA, RNA, CSF, and some biospecimens

    *Service provided*: data share, disease education, and biospecimen share

## 4.3 Data Platform for Alzheimer's Disease

### *4.3.1 Objective of Data Platform*

Compared with other health-related data platforms, current AD data platform mostly focuses on the data collection and access, and only a few services could be achieved by researchers. In order to facilitate the AD research and prevention, an idle AD data platform should satisfy the following objectives:

- Integrate various institutes' data, and supply authorized data access to enable the big data analysis for AD.
- Provide comprehensive data services to simple researchers' analysis, such as data management, data visualization, data statistics, data analysis tools, etc.
- Offer application programming interface (API) to promote AD's prevention.

According to the above objectives, we aim to construct a data platform framework for AD and specify its detailed services and technologies.

### *4.3.2 Framework of Data Platform*

The data platform consists of four modules, including entrance module, service module, data module, and computing module, which are demonstrated in Fig. 4.1. Users collect various types of data and upload them through entrance module under the agreement. Users that want to access the data should make a request and get the service under the permission of the committee. The functions of the four modules are described below:

- Entrance Module

    Entrance module is used to achieve the load balance and reverse proxy of various requests. It commutes with service module to orient users' http request to

# 4 Data Platform for Alzheimer's Disease

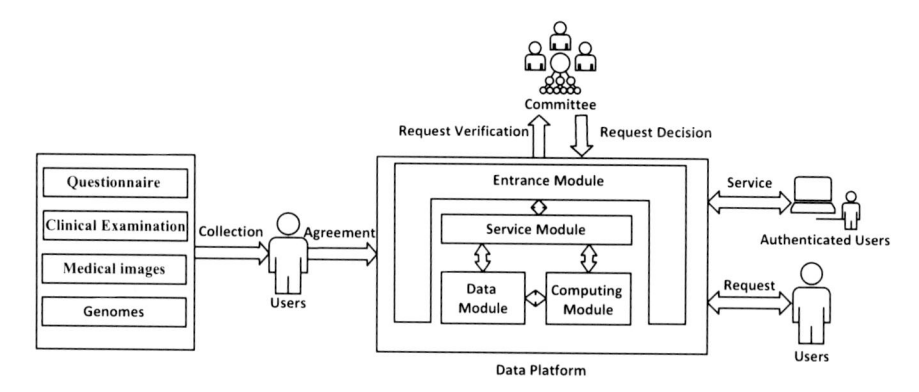

**Fig. 4.1** Framework of data platform. The platform consists of four modules: entrance module, service module, data module, and computing module. Entrance module receives various requests and commutes with service module to orient these requests to the corresponding service of a specific node in the server. Service module calls functions in data module and computes the module to accomplish the requests. Data module is responsible for data storage, and computing module handles the computation tasks

the corresponding service of a specific node in the server. Besides, it also identifies users' authority.

- Service Module

    Service module receives users' requests from entrance module and handles these requests by calling the corresponding service function in this module. Sometimes, it should also refer to the functions in data module and computing module to realize the service function. Thus, commutations would happen between service module and data module or computing module.

- Data Module

    Data module is responsible for organizing various types of data. It provides APIs for service module and computing module, such that these two modules could achieve data access. It should also balance the storage load.

- Computing Module

    Computing module is in charge of various computing operation. It integrates various computation functions, such as statistics, machine learning modeling, etc. Hence it needs to communicate with data module for data access and with service module for returning data computation results. The computation of load balance is also realized in this module.

## 4.4 Services of Data Platform

The data platform fulfills its objectives by various publicly available services, including data directory, data share, data management, data analysis support, etc. Figure 4.2 illustrates the composition of platform's services, and the detailed descriptions of these services will be introduce below.

### *4.4.1 Data Directory Service*

Data directory service aims to provide users the landscape of data stored in the data platform, such as data types, data size, etc. Besides, users could query these data information. The following data types should be contained in the data platform.

- Questionnaire data

   This type of data consists of various epidemiological investigation data, including socioeconomic, medical history, physical activity, and sleep questionnaire, food frequency questionnaire, environmental exposure, cardiovascular assessment, depression assessment, cognitive assessment, etc. This kind of data usually takes numeric or text format. By analyzing these data, risk factors could be determined to identify AD's prevention strategy or achieve AD screening.
- Clinical examination data

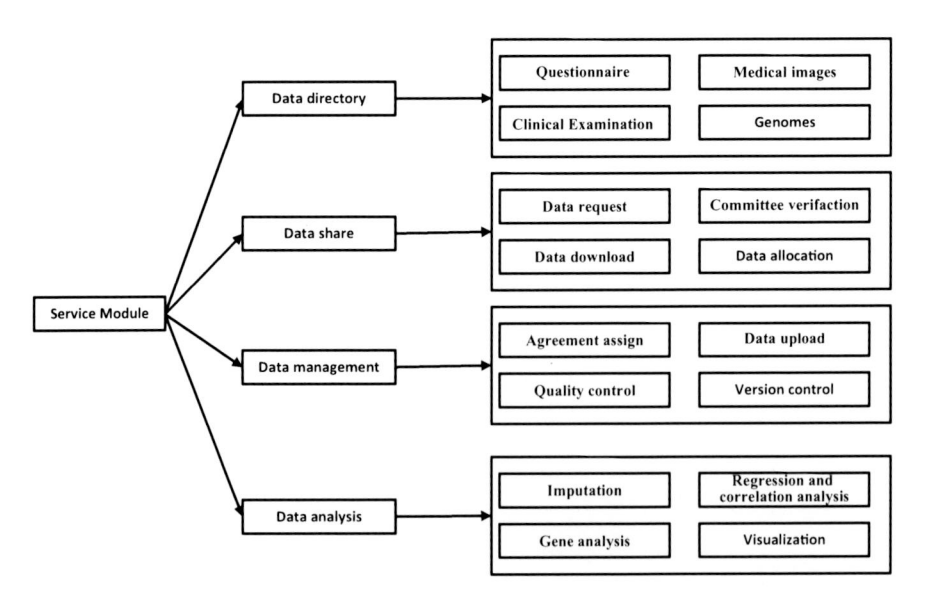

**Fig. 4.2** Services provided by the data platform. The service module contains four types of service, including data directory, data share, data management, and data analysis support

This type of data is obtained through various clinical examinations, including blood examination, urine examination, physical examination, etc. The format of these data is the same as questionnaire data, i.e., numeric or text. Biomarkers and risk factors could be derived from these examination data to help reveal AD's pathogenesis, identify AD's prevention strategy, or diagnose the disease.

- Medical images' data

  Medical images' data comprises of various images, such as magnetic resonance imaging (MRI), positron emission tomography (PET), electrocardiogram (ECG), etc. These data are usually used to derive biomarkers to help diagnose AD.

- Genomes' data

  Genomes' data is the genomes' information for each individual. These could be represented as sequences in database. Researchers analyze genomes to find biomarkers in order to reveal AD's pathogenesis.

### 4.4.2 Data Share Service

An important reason of integrating data from various institutes is to break the data access barrier and make big data available for researchers. Hence, data share service is a fundamental service.

Data are collected from different institutes, and privacies need to be protected. Thus, data access should obey an authorized procedure. First, researchers upload their access request, including types, sizes, plan of utilization, etc. Then the request will be transferred to the community for verification. Once the request gets improvement, a download link will be sent to the researchers for access. Besides, an agreement needs to be assigned by the requesters. The agreement indicates privacy protection, data share protocol, limit of data utilization, etc.

### 4.4.3 Data Management Service

Researchers usually collect their own data for analysis; therefore, they need to organize and manage these data, which is complicated and time-consuming. In order to simplify these processes, the platform provides a data management service for researchers to upload their data and conduct management.

The first concern needing to care about is the protection of each researcher' data, e.g., who can access these data, etc. The protection could be classified into four levels, including private, limited access controlled by the owner, limited access controlled by the platform community, and public. These levels indicate the data could only be accessed by the owner, accessed with the owner's permission, accessed with the platform community's permission, and accessed publicly, respectively. Before a researcher uses the platform to manage its data, an agreement needs to be assigned to illustrate the protection level, each responsibility, etc.

The data platform provides public interface for users to upload their data, and data directory service in Sect. 1.4.1 defines the data types. Besides, the interface should clarify data format for data storage. There have been some successful format including csv, plink, vcf, fasta, etc. They could be adopted by the data platform.

After uploading the data, users could revise the pretension level of their data and utilize platform's service to achieve data quality control, data analysis, etc. Quality control consists of data imputation, data quality verification, data collection process control, etc. Data analysis refers to generating data report by visualization, statistics analysis, data mining, etc.

Besides, the platform provides version control function for users' data. The user could choose to roll data, and create data labels, such that users could manage numerous amounts of data with frequent change, and the platform could still guarantee data's safety.

### 4.4.4 Data Analysis Support

Data analysis is a necessary strategy for researchers to achieve inspiring results. The platform offers comprehensive data analysis functions to support researchers' analyzing requirement. These functions include data imputation, machine learning modeling, genome analysis, and alignment, which are described below:

- Data imputation

    There may be missing values in the collected data due to various reasons, e.g., misoperation, noncooperation, etc. If analyzing the data with missing values directly, some bias may be introduced, resulting in unreliable outcomes. Imputing these missing values could make up these disadvantages to some extent. Therefore, the platform adopts some classical imputing methods, such as expectation-maximization (EM), k-nearest neighbors (KNN), least squares fitting (LLS), etc. Users could choose these imputation options to fill up missing values.
- Machine learning modeling

    Sometimes, researchers need to determine the relationships between various types of data or apply these data to make prediction. Then machine learning models should be established for these analyses. The platform incorporates diverse machine learning modeling strategies to fulfill this requirement, for example, correlation analysis for identifying relationships; logistic regression, SVM, or neural networks for predicting specific values; k-means or spectral clustering for unlabeled data clustering; principal component analysis (PCA) or locally linear embedding (LLE) for dimensionality reduction; etc.
- Genome analysis

Biomarkers could be derived from genomes; thus, genome analysis may be needed to achieve this function. The platform would integrate some basic genome analysis tools, such as sequence alignment, gene query, etc.

Data visualization could provide graphic representations for data and analysis results. Some underlying structures may be found through these visualized images. Thus, visualization is also an important analysis demand for researchers. The platform would provide the following visualization services.

- Frequency statistics visualization

  Users could check the visualized frequency statistics based on ages, gender, region, education levels, etc. They could also make filtering to the visualization according to specific attributes.

- Dimensionality reduction visualization

  Usually, a sample contains hundreds or thousands of attributes; thus, it is impossible to visualized the sample in a 2-D or 3-D graphic without dimensionality reduction. In order to achieve this type of visualization, the platform adopts linear and nonlinear dimensionality reduction methods to reduce the samples' dimension to 2 or 3, such that they could be visualized. Linear dimensionality reduction includes linear discriminant analysis (LDA), principal component analysis (PCA), etc. Nonlinear dimensionality reduction includes locally linear embedding (LLE), Laplacian eigenmaps, etc.

- Cohort study visualization

  Cohort study is a type of clinical study design. It is in the form of longitudinal study that aims to following a group of people for risk factor analysis. Therefore, the data changes with a time serial. The platform offers a dynamic cohort data visualization so that users could check the change of finite risk factors in the graphic format.

- Cross-sectional study visualization

  Cross-sectional study aims to provide data on the entire population under study. For this kind of data, the platform provides visualization option for researchers such that they could compare the data from different individuals by graphic format.

## 4.5 Implementation Technologies for Data Platform

In order to guarantee platform's safety and robust, some advanced technologies are adopted, which is illustrated in Fig. 4.3. The entrance module employs proxy system and load balance system, and the service module is implemented by Nginx, NodeJS, and Python. There are more complex issues to be concerned of when implementing data module and computing module; hence, the detailed technologies for these two modules are described below.

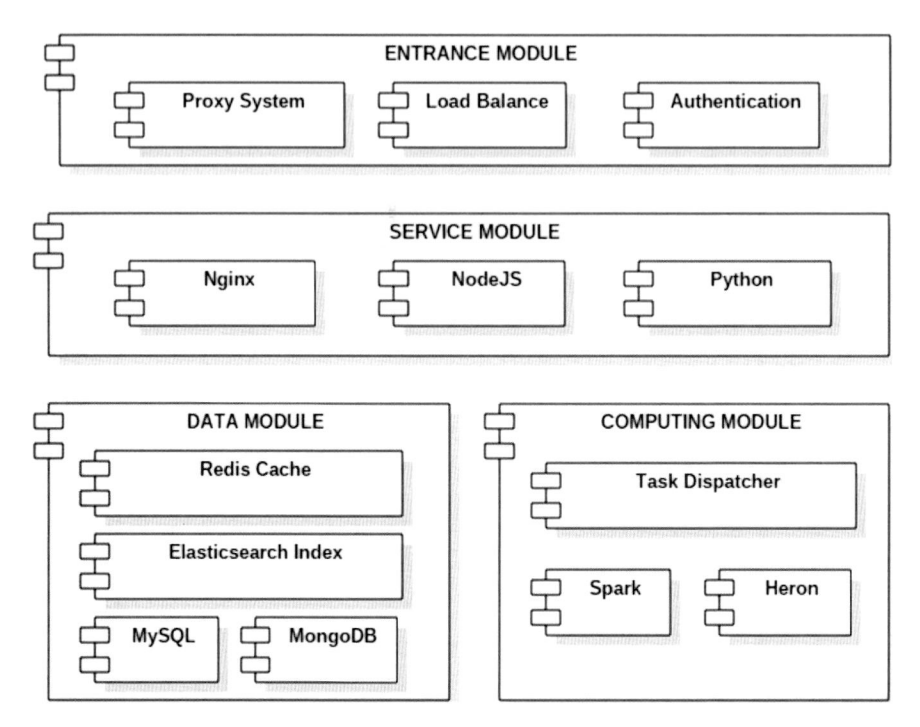

**Fig. 4.3** Technologies adopted by the data platform. The entrance module employs proxy system and load balance system, and the service module is implemented by Nginx, NodeJS, and Python. In data module, MySQL and MongoDB are used to store data, while Redis Cache and Elasticsearch index are employed to improve efficiency. Task Dispatcher, Spark, and Heron are applied in computing module to improve the computation efficient

### 4.5.1 Storage

There are two issues needing to be concerned, including:

1. The platform supports various types of data, and these data could not be treated equally due to different formats. There are mainly three data formats, i.e., structural data, such as tables; unstructured data, such as unstructured text and image data; and semi-structured data, which contain both data types.
2. The platform needs support for high synchronizing access, and distributed storage should be adopted due to huge amount of data.

Only using purely relational database to store these data is unrealistic. Therefore, the framework in 1.3.2 illustrates that the platform adopts a specific system to solve these problems.

The storage consists of four parts:

1. Cache layer:

Redis is applied to cache users' frequency data since it stores data in memory and supports numerous data types (including txt, number, row, dictionary, etc.).

This layer is used to accelerate user's data access and release the burden of storage system.

2. Index layer:

Elasticsearch is employed to provide index for txt files due to two reasons, one is that it can provide strong txt data index and composite index and the other is that it can accelerate the speed of index and release the burden of storage system.

3. Structural data storage:

The platform uses MySQL to store structural data. MySQL is the most well-known open-source relational database engine, which is stable, is low cost of training and establishment, and supports distributed database cluster with outstanding performance among open-source products.

4. Unstructured data storage:

The platform uses MongoDB to store unstructured data, including image data and document data. MongoDB supports master-slave distributed cluster storage and has strong querying function.

### 4.5.2 Computation

Computation module accomplishes all the computation task of the data platform. There are two issues that need to be concerned of for this module:

- Huge amount of computation tasks: computation technologies need to be adopted to improve the computing efficiency.
- Real-time demand: when utilizing the data analysis services, users may not wait for a long time for the results; this proposes real-time demand for computation module.

To tackle these two issues, the mixed computing framework is applied, including Spark and Heron. Spark is a streaming model-based platform, which is implemented by Java and provides multi-language interfaces. It is suitable for handling batch processing of big data and weak in real-time processing. Heron is a streaming model-based real-time platform, which could read and deal with huge amounts of data. All the temporal data analysis would be assigned to Heron. These two platforms are not independent. When requiring numeral amount of batch processing, Spark will be in charge.

Figure 4.4 illustrates the data flow in the computation module. Spark handles the whole data analysis for each day, which could be regarded as the preprocessing of data analysis. The newly coming data would be handled in Heron, and then the results would be combined with the results obtained by Spark.

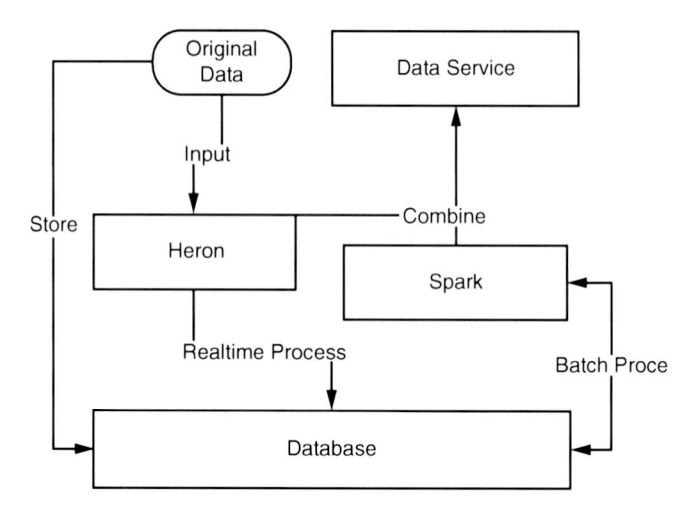

**Fig. 4.4** Data flow in the computation module. Spark handles the whole data analysis for each day. The newly coming data would be handled in Heron, and then the results would be combined with the results obtained by Spark

## 4.6 Prevention of AD Based on Data Platform

The objective of establishing data platform is to facilitate AD's research and prevention. In this section, concrete strategies for advancing AD's prevention are introduced, including how to determine risk factors, conduct cognitive assessment, and improve cognition.

### 4.6.1 Risk Factor Determination

One of the important tasks for prevention is to identify AD's risk factors to guide prevention strategies. This task is based on the analysis of epidemiological investigation data. By calling the data analysis support service, the relationships between risk factors and the diseases could be determined.

Given a set of risk factors, the following analysis support could be applied for risk factor analysis. Student's t-test and $\chi^2$ test are used to evaluate the risk factors' differences in midlife characteristics by dementia status; logistic regression analyses are used to validate the CAIDE risk score, which is calculated by the Kaplan-Meier method; C statistic is applied to evaluate the dichotomous outcome of ever/never dementia, where C statistic is the area under ROC curve; Cox proportional hazard models could be used for prediction modeling; partial least squares regression could be employed for risk factors selection in predicting AD [27].

## 4.6.2 Cognitive Assessment

Cognitive assessment is a technique used to assess the logical or reasoning abilities of human brains. It is critical in AD's prevention as it could be used for (1) screening for cognitive impairment, (2) differential diagnosis of cause, and (3) rating of severity of disorder or monitoring disease progression [28].

The assessments are mostly focused on attention, memory, visuospatial skills, and executive function abilities. The common methods for cognitive assessment are task-oriented assessment, which ask participants to complete a series of tasks like logical puzzles, matching numbers, etc. that necessitate the involvement of cognitive skills. Some classical tasks includes Prevention and Early Intervention Program for Psychoses (PEPP) cognitive assessment battery, Psychology Experiment Building Language (PEBL), Bhatia's Battery for Performance Test of Intelligence, neurocognitive test battery, cognitive drug research (CDR) computerized assessment system, etc. [29]

Some assessment tools have been developed based on these assessment tasks. The platform provides APIs to integrate these tools to facilitate AD's prevention. Some successful tools are described below:

- Alzheimer's Disease Pocketcard

    Alzheimer's Disease Pocketcard app can manage AD with confidence. It could help physicians and other healthcare professionals to take care of AD patients. Its highlights include (1) top ten signs of Alzheimer's disease; (2) latest information on detection, diagnosis, and management of Alzheimer's disease; (3) interactive tools to assess cognition and function, including the Mini-Cog, clock-drawing test, Saint Louis University Mental Status Exam (SLUMS), Functional Activities Questionnaire (FAQ), etc.; (4) annual wellness visit algorithm to help clinicians assess cognition more efficiently; (5) current diagnostic criteria, including the DSM-5 and the updated diagnostic criteria and guidelines for Alzheimer's disease from the National Institute on Aging and the Alzheimer's Association; and (6) education/support packets (PDF brochures) from the Alzheimer's Association that can be e-mailed directly to patients and caregivers.
- ANU Alzheimer's Disease Risk Index (ANU-ADRI)

    ANU-ADRI is an effective evidence-based tool to assess AD risk factors of people after 60 years old; it can provide personalized evaluation and exposure risk factors. It can help people know their current risk profile and areas, to reduce the risk of Alzheimer's disease in these areas. In addition, it can help the doctor to record their risk profile for the next consultation. ANU-ADRI can also be used for research projects to assess in reducing Alzheimer's risk [30].
- The CAIDE Risk Score (mobile application) App

    CAIDE dementia risk score app is a validated tool to predict the risk of dementia in his later years (20 years later) by involving age, education level, high blood pressure, high cholesterol, obesity, and lack of exercise. It can help users to reduce the modifiable risk factors and deferred cognitive impairment and dementia. Users could test their own personal risk and check the guidance for the modifiable risk. The app would recommend to consult a professional doctor if necessary. It also allows doctors to discuss preventive measures and monitor the reduction of risk [31].

### 4.6.3 Prevention Strategies and Cognitive Improvement

Cognitive assessment is to evaluate the cognitive status, while risk factor determination is to find the AD-related risk factors. Both strategies could provide guidance for AD prevention and improvement. The platform provides a section to maintain this guidance. Some typical guidance are introduced as follows.

Lifestyle can reduce the incidence of AD which contains physical exercise, cognitive ability, level of education, and social competence. The effect of aerobic exercise may come from the improvement of cardiovascular health and cardiovascular health. In the mouse models, environmental enrichment, including repeated exposure to novelty, has been shown to reduce amyloid burden associated with neuroanatomical and behavioral defects. Powerful social participation may help to reduce the likelihood of elderly peoples' suffering from dementia including AD. It may also help to slow down the development of symptoms [32].

These guidances may not be easy to access from the website; thus, the platform provides APIs to allow mobile phones to access these information. Some successful apps for cognitive improvement have been developed, e.g., ADcope. ADcope contains five modules: the first one is memory wallet, which consists of 30 images or sentences about the familiar family, places, and events; the second one is calendar, which is used to remind patients of daily activities; the third one is NFC tags, which are placed on drawers, doors, etc.; by clicking the tag, the app will display the contents of the drawer or the room; the fourth one is audio-assisted memory training module, which could train the patient's memory by repeating biographical information and asking questions about the information; and the last one is spaced retrieval exercise module, which is used to enhance patient's memory ability by two-phase exercise, assessing, and training [33].

The strategies for improving cognitive usually require users to accomplish serials of activities. How to monitor the activities is an issue needing to concern. Fortunately, many smart devices have been developed to monitor human bodies' physical or motion information. Some typical smart devices are described below:

1. Smart bracelet

    Smart bracelet uses wrist smart devices to measure the heart rate and obtain simplified heart rate zone continuously and automatically. Take the Fitbit smart bracelet as an example; it integrates training mode to record exercise, and people could view the details and summary of exercise data in the smart phone in real time. Besides Fitbit smart bracelet can record steps, distance, calories book, numbers of stairs, and the active time. It could also automatically monitor the sleep hours and sleep quality. Users could set a target, record diet, view progress, and analyze trends by mobile phone.

2. Smart clothes

   Smart clothes use flexible sensor technology to monitor the motion of the heart rate, breathing, and the main muscle groups of the body surface in real time. Take BodyPlus as an example; people can develop the most suitable fitness plan for their own according to their situation. BodyPlus equipped with powerful data analysis tools could sum up and analyze people's training and generate an analysis report for the heart and lung function and muscle stimulation situation.

3. Intelligent headset

   When users exercise with a headphone, the embedded motion sensor can monitor the body's temperature, perspiration, heart rate, and other indicators. Besides, earplugs' base frame is also equipped with acceleration sensors, making it achieve more precise motion data.

4. Intelligent insole

   Intelligent insole could be used to monitor human's respiration, heart rate, exercise, sleep, gait, and other information continuously.

5. Smart mattress

   Smart mattress could measure users' breathing and heart rate during sleep and send the collected data to smart phones wirelessly. The corresponding mobile application could show the depth of sleep, snoring time, etc., and assess the quality of people's sleep and give suggestions for improvement.

These smart devices need to communicate with smart phone for data processing and visualization. Some data process may be too complex to be run on the phones. The platform would provide APIs for these smart devices to upload their data for analysis. Through the data service module, the more comprehensive data analysis functions could be obtained.

## 4.7 Research of AD Based on Data Platform

The platform incorporates comprehensive data analysis methods to support AD's research. Some classical cases for these analysis methods' application are introduced below, including biomarkers extracted from medical images, diagnosis assistant, AD prediction, etc.

### *4.7.1 Biomarkers Extracted from Medical Images*

Various types of medical images have been integrated in the platform, including magnetic resonance imaging (MRI), positron emission tomography (PET), etc. These images contain specific topologies, and biomarkers extracted from

AD/MCI's MRI image and PET image have been proven to be able to aid diagnosis [36].

Currently, medical images' biomarkers are mostly extracted manually, which requires professional background knowledge. Some doctors lacking in AD image knowledge may not diagnose AD accurately. These wrong diagnoses of AD would cause patients' unnecessary fear, despair, or even discrimination. Hereby, biomarker extraction becomes an important task for AD research and contributes significantly to the diagnosis and pathological study of AD.

The platform provides back-propagation (BP) neural network method for biomarkers derived from medical images. This method is one of the most famous machine learning modeling methods to recognize and classify features from medical image and has been widely used in the diagnosis systems of lung cancer. Users could utilize BP neural network method provided by the platform to extract biomarkers from MRI and PET. By establishing a multilayer feedforward neural network, the method uses numeral numbers of image data as training set and adjusts parameters based on mean square error to achieve biomarker extraction.

Data platform also supports convolutional neural network (CNN) for image analysis. CNN is a deep learning method, which adds a new step which imitates the way of human brain's processing signal for feature learning. CNN uses convolutional layer and dimension reduction layer to make the computer to extract biomarkers close to medical expert, which could improve the accuracy of extracting biomarkers from AD image data.

### 4.7.2 Diagnosis Assistant

Currently, AD diagnosis depends on the pathological diagnosis. An issue emerges that AD couldn't be definitely diagnosed before patients' death. Research shows that the misdiagnosis rate of AD range from 27 to 57%. Therefore, how to improve the accuracy of AD diagnosis is the most crucial problem of AD-related research.

Actually, diagnosis could be converted into machine learning classification problem, which requires two stages: training and classification. The former one is to analyze data set and establish classification models, and the latter one is to apply the new sample to the model and output the prediction value. In short, the whole process could be regarded as compressing patients' data into a number associated with AD so that the patients could be classified whether or not the patients suffer from AD [37].

The platform integrates various machine learning algorithms for AD diagnosis assistant, including support vector machine (SVM), relevance vector machine (RVM), back-propagation neural network, etc. The following introduce how to apply these algorithms on different types of data to make a diagnosis:

- *Diagnosis based on questionnaire data*

  For questionnaire data, random forest algorithm could be applied. Take Alzheimer's Disease Assessment Scale-Cognitive Subscale (ADAS-Cog) as an example; random forest is employed to generate the scale of ADAS-weighted tree. According to the ADAS-weighted tree, we can estimate whether the patient is normal, AD, or MCI. Compared with magnetic resonance imaging (MRI) or extraction of cerebrospinal fluid biomarker measurement, ADAS-weighted tree is considered to be more economical and useful. Moreover, the ADAS tree method won't cause great influence to the investigators' health.

- *Diagnosis based on image data*

  Support vector machine (SVM) is a widely used method for image-based diagnosis. By using biomarker extracted from MRI or PET image data, a classifier could be established based on SVM to classify the biomarker data and assign each sample a value as a label. These labels could be regarded as the diagnosis results.

  SVM uses kernel function to transfer the input space into a high-dimensional space, and then an optimal classified hyperplane is determined. Sometimes, dimension reduction is applied to reduce the computation complex before establishing SVM classifier model, which means some information may be discarded. This could introduce some errors. Hereby, the platform provides penalized logistic regression and coordinate-wise optimization methods to overcome the disadvantage of dimension reduction.

  Furthermore, relevance vector machine (RVM) algorithm is provided for AD classification, which is based on sparse Bayesian framework. RVM first transfer the input data into a high-dimensional space by choosing a proper kernel function such as Gaussian kernel function, Laplace kernel function, polynomial kernel function, etc. Among which, Gaussian kernel is the most widely used kernel. Then RVM needs to initialize the hyper-parameter and obtain the ideal model of the classifier by iteration. Compared to SVM algorithm that is based on structural risk minimization, RVM may have a stronger generalization ability.

- *Diagnosis based on multimode data*

  Previous methods are based on single type of data. Sometimes, multiple types of data are available. Combining various types of data to make a diagnosis could achieve more accuracy results [38], for example, any two types or all types of questionnaire data (MMSE, ADAS-Cog, CDR), image data (MRI, PET), gene data, and biomarker data (CSF) could be combined. Therefore, the platform provides options for multimode data analysis.

  One method is to integrate both MRI image data and MMSE scale to make a diagnosis. By building resting-state brain functional networks, we can extract the abnormal network nodes' properties and train the SVM classifier based on voxel-based morphometry (VBM).

- Another method is to combine image data (FDG-PET image, MR scan image), extracted biomarker data of cerebral spinal fluid (CSF), and questionnaire data (MMSE, ADAS-Cog) to make a classification of AD, MCI, and NC. This is also a SVM-based method. At first, the algorithm processes feature extraction and feature selection from image data and calculates the kernel matrix based on selected feature. For CSF data, the algorithm directly calculates the kernel matrix without feature selection. Then the method combines all kernel matrixes (MRI, PET, CSF) to process classification by using SVM algorithm. Comprehensive studies show that this method could achieve higher accuracy than the other methods based on single type of data.

### 4.7.3 AD Prediction

Another important AD research is to make prediction about the progress of the disease, e.g., the onset time, disease exacerbation, transformation of disease symptoms, etc. This relies regression analysis of cohort clinical data. Thanks to the development of epidemiological investigation, many cohort data have been collected, which make the AD prediction possible [39].

The prediction could be done on single type of data, but this may not achieve inspired performance. Therefore, the platform supports a multimode prediction method based on combined different kinds of data. This method not only makes regression analysis from collected clinical data but also combines with the mentioned classification method in the previous chapter. Besides, we use the corresponding multimodal support vector regression (SVR) algorithm as the basis. This method not only predicts clinical data through the regression method but also predicts the 2-year changes of MMSE and ADAS-Cog scores to help validate the prediction results.

## 4.8 Conclusion

Data share is an important strategy to advance AD area. Therefore, data platform is designed to integrate data from various institutes to enable big data analysis, which could accelerate AD's research and prevention. Besides, comprehensive services are provided to simplify searchers' analysis task for data. Prevention and research cases based on data platform illustrate the importance of the data platform. Those who devote to AD area should turn to the data platform for data share and analysis support.

**Acknowledgments** This study was supported partially by Anhui Provincial Key Technologies R&D Program under Grant No. 1704e1002221, the National Natural Science Foundation of China (NSFC) under Grant No. 71661167004 and the Programme of Introducing Talents of Discipline to Universities ("111 Program") under Grant No. B14025.

## References

1. Hebert LE, Weuve J, Scherr PA, Evans DA (2013) Alzheimer disease in the United States (2010–2050) estimated using the 2010 census. Neurology 80(19):1778–1783
2. Prince M, Wimo A, Guerchet M, Ali GC, Wu YT, Prina M. (2015) The global impact of dementia: an analysis of prevalence, incidence, cost and trends. World Alzheimer Report
3. Wimo A, Jönsson L, Bond J, Prince M, Winblad B, International AD (2013) The worldwide economic impact of dementia 2010. Alzheimers Dement 9(1):1–1
4. Seshadri S, Beiser A, Kelly-Hayes M, Kase CS, Au R, Kannel WB, Wolf PA (2006) The lifetime risk of stroke estimates from the Framingham study. Stroke 37(2):345–350
5. Mahmood SS, Levy D, Vasan RS, Wang TJ (2014) The Framingham heart study and the epidemiology of cardiovascular disease: a historical perspective. Lancet 383(9921):999–1008
6. Alzheimer's A (2015) 2015 Alzheimer's disease facts and figures. Alzheimers Dement 11 (3):332
7. Shimmer, Wireless Sensing Solutions for werable applications, online at: http://shimmer-research.com/
8. Souillard-Mandar W, Davis R, Rudin C, Au R, Libon DJ, Swenson R, Price CC, Lamar M, Penney DL (2016) Learning classification models of cognitive conditions from subtle behaviors in the digital Clock Drawing Test. Mach Learn 102(3):393–441
9. Zlokovic BV (2011) Neurovascular pathways to neurodegeneration in Alzheimer's disease and other disorders. Nat Rev Neurosci 12(12):723–738
10. Honig LS, Kukull W, Mayeux R (2005) Atherosclerosis and AD analysis of data from the US National Alzheimer's coordinating center[J]. Neurology 64(3):494–500
11. Shaw LM, Vanderstichele H, Knapik-Czajka M et al (2009) Cerebrospinal fluid biomarker signature in Alzheimer's disease neuroimaging initiative subjects[J]. Ann Neurol 65 (4):403–413
12. Jack CR, Bernstein MA, Fox NC et al (2008) The Alzheimer's disease neuroimaging initiative (ADNI): MRI methods[J]. J Magn Reson Imaging 27(4):685–691
13. Hodes RJ, Buckholtz N (2016) Accelerating medicines partnership: Alzheimer's disease (AMP-AD) knowledge portal aids Alzheimer's drug discovery through open data sharing [J]. Expert Opin Ther Targets
14. Simmons A, Westman E, Muehlboeck S et al (2011) The AddNeuroMed framework for multi-centre MRI assessment of Alzheimer's disease: experience from the first 24 months[J]. Int J Geriatr Psychiatry 26(1):75–82
15. Lovestone S, Francis P, Kloszewska I et al (2009) AddNeuroMed—the European collaboration for the discovery of novel biomarkers for Alzheimer's disease[J]. Ann N Y Acad Sci 1180 (1):36–46
16. Partch AB, Laufer D, Valladares O et al (2015) Nia genetics of Alzheimer's disease data storage site (NIAGADS): 2015 update[J]. Alzheimers Dement 11(7):P362
17. Pérez-Cano R, Vranckx JJ, Lasso JM et al (2012) Prospective trial of adipose-derived regenerative cell (ADRC)-enriched fat grafting for partial mastectomy defects: the RESTORE-2 trial[J]. Eur J Surg Oncol (EJSO) 38(5):382–389
18. Thal LJ (2004) The Alzheimer's disease cooperative study in 2004.[J]. Alzheimer Dis Assoc Disord 18(4):183–185
19. Zhao Y, Hu Y, Smith JP et al (2014) Cohort profile: the China health and retirement longitudinal study (CHARLS).[J]. Int J Epidemiol 43(1):61
20. Cerami E, Gao J, Dogrusoz U et al (2012) The cBio cancer genomics portal: an open platform for exploring multidimensional cancer genomics data[J]. Cancer Discov 2(5):401–404
21. CommonMind Consortium Launched as a Public-Private Effort to Generate and Broadly Share Molecular Data on Neuropsychiatric Disease[J] (2012) Biomedical Market Newsletter
22. Day K, Mcguire L, Anderson L (2009) The CDC healthy brain initiative: public health and cognitive impairment[J]. Generations 33(1):11–17

23. States A (2013) The healthy brain initiative: the public health road map for state and national partnerships, 2013–2018[J]. Aging/physiology/united States
24. Jiang J (2012) Discussion on the information Organization of the Scientific Data Sharing Platform-Take the National Scientific Data Sharing Platform for population and health as an example[J]. J Inf Resour Manag
25. Centers for Disease Control and Prevention (2014) Health indicators warehouse[J]. Obesity in children and adolescents aged. 2–19
26. Parkinson J (2008) Common neurological disorders: Parkinson's disease[J]. Br J Healthc Assistants January
27. Exalto LG, Quesenberry CP, Barnes D, Kivipelto M, Biessels GJ, Whitmer RA (2014) Midlife risk score for the prediction of dementia four decades later. Alzheimer's Dement: J Alzheimer's Assoc 10:562–570. doi:10.1016/j.jalz.2013.05.1772
28. Woodford H, George J (2007) Cognitive assessment in the elderly: a review of clinical methods. QJM 100:469–484
29. Singh M, Sachdeva S (2014) Cognitive assessment techniques. Int J Inf Technol Knowl Manag 7(2):108–118
30. Anstey KJ et al (2014) A self-report risk index to predict occurrence of dementia in three independent cohorts of older adults. The ANU-ADRI. PLoS One 9:e86141
31. Sindi S, Calov E, Fokkens J et al (2015) The CAIDE dementia risk score app: the development of an evidence-based mobile application to predict the risk of dementia. Alzheimers Dement 1:328–333
32. Selkoe DJ (2012) Preventing Alzheimer's disease. Science 337(6101):1488–1492
33. Zmily A, Mowafi Y, Mashal E (2014) Study of the usability of spaced retrieval exercise using mobile devices for Alzheimer's disease rehabilitation. JMIR mHealth uHealth 2(3):e31
34. Bloom GS (2014) Amyloid-$\beta$ and tau: the trigger and bullet in Alzheimer disease pathogenesis. [J]. JAMA Neurol 71(4):505–508
35. Pastor P, Roe CM, Villegas A et al (2003) Apolipoprotein E$\varepsilon$4 modifies Alzheimer's disease onset in an E280A PS1 kindred[J]. Ann Neurol 54(2):163–169
36. Gudbjartsson H, Patz S (1995) The Rician distribution of noisy MRI data[J]. Magn Reson Med 34(6):910
37. Escudero J, Ifeachor E, Zajicek JP et al (2013) Machine learning-based method for personalized and cost-effective detection of Alzheimer's disease[J]. IEEE Trans Biomed Eng 60 (1):164–168
38. Zhang D, Wang Y, Zhou L et al (2011) Multimodal classification of Alzheimer's disease and mild cognitive impairment[J]. NeuroImage 55(3):856–867
39. Hinrichs C, Singh V, Xu G et al (2011) Predictive markers for AD in a multi-modality framework: an analysis of MCI progression in the ADNI population[J]. NeuroImage 55 (2):574–589

# Chapter 5
# Data Analysis for Gut Microbiota and Health

### Xingpeng Jiang and Xiaohua Hu

**Abstract** In recent years, data mining and analysis of high-throughput sequencing of microbiomes and metagenomic data enable researchers to discover biological knowledge by characterizing the composition and variation of species across environmental samples and to accumulate a huge amount of data, making it feasible to infer the complex principle of species interactions. The interactions of microbes in a microbial community play an important role in microbial ecological system. Data mining provides diverse approachs to identify the correlations between disease and microbes and how microbial species coexist and interact in a host-associated or natural environment. This is not only important to advance basic microbiology science and other related fields but also important to understand the impacts of microbial communities on human health and diseases.

**Keywords** Microbiome • Data mining • Data analysis • Microbiota • Microbes • Diseases

## 5.1 Introduction

There are more and more evidences to confirm that human "microbiome" – microbes living in intimate association with us – forms a vital part of our biology and plays an important role in both health and sickness [1]. Metagenomics methods which sequence DNA without directly identifying [2] which organisms they come from and 16s rRNA sequencing [3] which sequence tag DNA for identifying the composition of organisms are two basic way of microbiome analysis.

Recently, huge amount of data are generated from plenty of microbiome projects such as Human Microbiome Project (HMP) [4, 5] and Metagenomics of the Human Intestinal Tract (MetaHIT) [6]. These datasets provide great opportunities to study

---

X. Jiang (✉)
School of Computer, Central China Normal University, Wuhan, Hubei 430079, China
e-mail: xpjiang@mail.ccnu.edu.cn

X. Hu
School of Computer, Central China Normal University, Wuhan, Hubei 430079, China

College of Computing & Informatics, Drexel University, Philadelphia, PA 19104, USA

the unknown world of microbes. Analyzing and mining these data will help us to better understand the function and structure of microbial community of the human body, thus the relationships to our health [7, 8].

However, the huge data volume, the complexity of microbial community, and the intricate data properties have introduced challenges for microbiome data analysis and mining [9, 10]. Bioinformaticians including computer scientist, mathematician, and microbiologist work together to develop computational approaches to tackle these challenging issues, roughly focusing on the following computational tasks: (1) dimension reduction and visualization approaches to explore and visualize microbiome data, (2) statistical methods to infer true correlations and relationships between microbes and diseases, (3) computational methods to identify and extract microbial interactions from microbiome datasets, and (4) dynamic modeling and time series analysis to model the ecological system in a holistic way.

Metagenomic data analysis is a timely topic; there is great need for better algorithms to analyze complex microbiome datasets. These efforts undoubtedly will lead to biological insights on how microbes impact human health. We will breifly introduce the current advances in four aspects mentioned above in microbiome data analysis and mining.

## 5.2 Dimension Reduction and Pattern Identification

After the preprocessing of the metagenomic data, DNA of metagenomic or 16s rRNA sequencing technologies could be summarized by metagenomic profiles [11] which summarize the abundance of functional or taxonomic categorizations in metagenomic sequences. A metagenomic profile matrix typically has hundreds of metabolic pathways, thousands of species or tens of thousands of protein families [12]. Machine learning and multivariate statistics have been employed on the profile matrix to explore and extract the complex patterns and correlations [13]. After dimension reduction, metagenomic profiles are usually represented by several "components" which may facilitate biological interpretation and discovery [14].

For example, PCA has been used frequently in metagenomic profiles to characterize the relationship of metagenomic samples [15]. Another method – MDS – which is based on the dissimilarities of data instead of similarity in PCA has been adopted as a standard technology for visualizing the taxonomic relationships in microbial communities [15]. Recently, a nonnegative matrix factorization (NMF) framework has been used in analyzing metagenomic profiles to gain a different and complementary perspective on relationships between functions, environment, and biogeography of global ocean and soil environment [16–18].

Microbiome datasets can be represented by metabolic paths, taxonomic assignment, or gene families [19]. To integrate information from multiple views, data integration approaches can be used to combine multi-view information simultaneously to obtain a comprehensive view which reveals the underlying data structure shared by multiple views [20]. A novel variant of symmetric nonnegative matrix factorization (SNMF) [21], called Laplacian regularized joint symmetric nonnegative matrix factorization (LJ-SNMF) has been proposed for this purpose. We conduct extensive experiments on several realistic datasets including Human Microbiome Project (HMP) data [4, 5]. The experimental results show that the proposed method outperforms other variants of NMF, which suggests the potential application of LJ-SNMF in clustering multi-view datasets.

Furthermore, linear correlation or regression methods are also employed to investigate the relationships among taxa or functions and their relationships to existing environmental or physiological data (metadata) such as Pearson Correlation and Canonical Correlation Analysis (CCA) [22]. CCA has been proposed for investigating the linear relationships of environmental factors and functional categorizations in global ocean [23].

The vast majority of methods employed in current metagenomics analysis are under the hypothesis that structures and relationships in a microbial community are linear. However, the interactions among microbiota are most likely nonlinear, and the mathematical spaces of microbiota are most likely in a manifold [24] or probabilistic space [25, 26] instead of Euclidean space. We could visualize and explore these structures using only several components which are the intrinsic dimensions discovered by manifold and probabilistic models. This provides a mechanistic understanding of how a microbial community is generated by probabilistic mixing of microbial components as well as a powerful tool for exploring the temporal dynamics of microbiome composition.

Finally, many kinds of nonlinear relationships such as taxa-taxa patterns and function-environment correlations could be investigated using the nonlinear statistical methods. We summarize these steps in a computational framework (see Fig. 5.1). The computational framework is based on our current understanding of metagenomic data, and we will integrate the advanced nonlinear dimension reduction methods and statistical methods to discover novel relationships.

## 5.3 Relationship and Correlation

Another important problem in microbiome analysis is to identify the biomarkers (i.e., bacterial taxa, microbial genes, or pathways) that are associated with disease, where the microbiome data are summarized as the composition of the bacterial taxa, protein families, or metabolic pathways at different levels [27]. To discover biomarkers for diseases or environmental factors, the most common approaches focus on regression techniques incorporating the complex interaction patterns among species (or gene functions). We have developed a new regression framework called "manifold-constrained regularization" (McRe) [28], which inherits the strength of manifold embedding for regularization of linear regression. This method can incorporate species interaction network as prior information to infer novel relationships.

Several studies consider the regression analysis of microbiome compositional data, where the goal is to identify the biomarkers that are associated with a continuous response such as the body mass index (BMI) [9]. Compositional data are strictly positive and multivariate that are constrained to have a unit sum. Lin

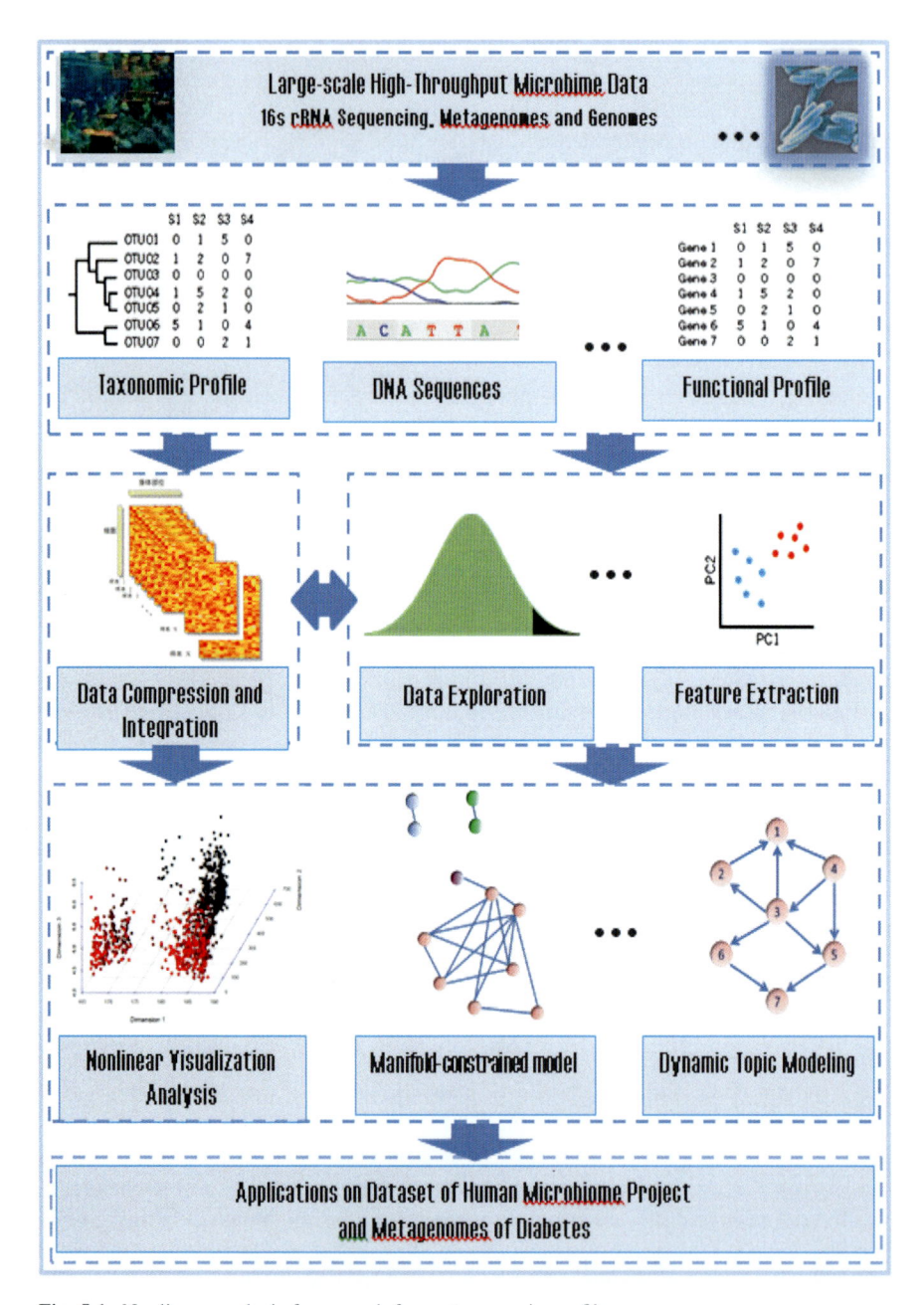

**Fig. 5.1** Nonlinear analysis framework for metagenomic profiles

et al. [29] proposed a variable selection procedure for such models in high-dimensional settings and derived the weak oracle property of the resulting estimates [29]. Shi et al. [30] proposed a penalized estimation procedure for estimating the regression coefficients and for selecting variables under the linear constraints which is developed [30]. This provides valid confidence intervals of the regression coefficients and can be used to obtain the p-values which could be used to measure statistical significance [30]. Randolph et al. have formulated a family of regression models that naturally extends the dimension-reduced graphical explorations common to microbiome studies; the method could be viewed as a penalized version of the low-dimensional linear model for compositions [31].

## 5.4 Networking Microbiome

A network perspective provides unprecedented opportunities for integrating and analyzing big microbiome data for studying the structure and the function of microbial communities [32]. A microbial interaction network (MIN, e.g., species-species interaction network) shapes the structure of a microbial community and forms its ecosystem function and principle, i.e., the regulation of microbe-mediated biogeochemical processes [33]. Deciphering interspecies interaction is challenging in the wet lab due to the difficulties of coculture experiments and the complicated patterns of species interactions [34]. The knowledge of these small-scale microbial interactions such as pairwise competitions is often distributed widely in various media including PubMed literatures, biological databases, Wikipedia documents, etc., making it difficult to integrate and analyze [35]. Researchers have started to infer pairwise interspecies interactions such as competitive and cooperative interactions leveraging to heterogeneous microbial data including metagenomes, microbial genomes, and literature data. These efforts have facilitated the discovery of previously unknown principles of MIN, verified the consistency, and resolved the contradiction of the application of macroscopic ecological theory in microscopic ecology [36].

Species interact in a complex style with many types of interactions unknown. Previous works on species inference based on metabolic methods are based on the following two approaches. Bornstein proposed a computational method for inferring pairwise interactions from reconstructed metabolic network of species with whole-genome sequences available publicly [37]. The method can identify pairwise competitive and cooperative interactions. Another way is using *flux balance analysis* (FBA) models [38] to infer species interaction when the metabolic model of a species (or strain) is available [39].

More than 100 genome-scale metabolic network models were published. Constraint-based modeling (CBM) was already used for the inference of three potential interactions [40]: negative, where two species compete for shared resources; positive, where metabolites produced by one species are consumed by another producing a synergistic co-growth benefit; and neutral, where co-growth has no net effect. By using the FBA simulation community metabolic network, we can find key enzymes and reactions in the metabolic network, thus acting as a potential environmental and physiological fingerprint. In a two species system, the CBM solver aims to explore the type of interactions by comparing the total biomass

**Fig. 5.2** A constraint-based modeling to model pairwise interaction

$$v_{BM,AB} = max \sum_{m \in \{A,B\}} v_{BM,m} \quad (1)$$

Subject to:

$$SV = 0$$

$$v_{i,min} \leq v_i \leq v_{i,max}$$

$$v_i \in V$$

production rate (denoted AB) in the pairwise system to the sum of corresponding individual rates recorded in their individual growth (denoted A+B). The CBM model is defined in Fig. 5.2, where $^V$BM,m is the maximal biomass production rate in a system m, corresponding to species A and B. When AB &lt;&lt; A+B, A and B have a competitive relationship.

## 5.5 Dynamics and Time Series Analysis

Microbial abundance dynamics along the time axis can be used to explore complex interactions among microorganisms [41]. It is important to use time series data for understanding the structure and function of a microbial community and its dynamic characteristics with the perturbations of the external environment and physiology [42]. Current studies usually use time sequence similarity [43], or clustering time series data for discover dynamic microbial interactions; these methods often do not take the full advantage of the time sequences. Thus the interactions among microorganisms cannot be accurately predicted. We have explored a vector autoregression (VAR) model [44] to lift the limitations of traditional methods. VAR models and interaction inference: Due to the high-dimensional nature of microbiomics data, the number of samples is far greater than the number of microorganisms; direct interaction inference by VAR is not feasible. In our previous studies, we have designed several graph regularization-based VAR (GVAR) methods for analyzing the human microbiome. We found that our approach improves the modeling performance significantly on several microbiome dataset. The experimental results indicate that graph regularization achieves better performance than other sparse VAR model based on elastic net regularization. However, the interpretation of the inference results is hard and far from complete. Furthermore, graph regularization, despite a classic manifold regularization method, suffers some problems because of its weak extrapolating ability. A novel regularization – Hessian regularization [45] – which fits the data perfectly and extrapolates nicely to unseen data will be utilized to overcome the issue.

In the future, state-space model [46] or probabilistic Boolean network model [47] could be used for modeling large-scale microbiome data for application. We will extend these methods by integrating specific information of the microbiomics data. The state-space model is a powerful method for simulation of dynamical systems, and it is widely used in engineering control systems which is a dynamic time-domain model to imply time as the independent variable. It is possible to extend the state-space model by considering the species delay in the regulatory network of relationships, not just describe the level of species richness' impact on the internal state, and assume that the internal state can independently evolve. Species with time delay regulatory network of relationships are better suitable for microbial interactions, because the regulation between microorganisms is often a slow process with delay, rather than an instantaneous process.

## 5.6 Conclusion

The data from Human Microbiome Project (HMP) [4, 5], which includes more than 5000 samples with profiles of hundreds of taxonomic or functional categorizations, are constructed from 15 or 18 distinct body sites of 242 individuals. Methodological development is still in its infancy for effectively analyzing and mining the data. Many microbiome dataset are also from various studies focusing on disease, diets, and other investigations. These data have created a great opportunity for understanding and also a tremendous computational and theoretical challenge. There is a great need to develop novel mathematical and computational methods for finding nonlinear signal and patterns in human-associated microbial metagenomes.

The identification of complex structures and patterns of microbial communities is at the essential part of studies in microbial ecology. The expected method helps shed light on discovering the complex relationships among microbes. In the future, nonlinear methods should be considered as an important tool in analyzing metagenomics, not only because microbial function can be viewed at multi-scales, from individual genomes to communities to global cycles, but also the complex interaction across scales.

## References

1. Shreiner AB, Kao JY, Young VB (2015) The gut microbiome in health and in disease. Curr Opin Gastroenterol 31(1):69–75
2. Comparative metagenomics of microbial communities. Science. [Online]. Available: http://science.sciencemag.org/content/308/5721/554. Accessed 04 Feb 2017
3. Woese CR, Kandler O, Wheelis ML (1990) Towards a natural system of organisms: proposal for the domains archaea, bacteria, and Eucarya. PNAS 87(12):4576–4579
4. T. H. M. P. Consortium (2012) Structure, function and diversity of the healthy human microbiome. Nature 486(7402):207–214
5. Turnbaugh PJ, Ley RE, Hamady M, Fraser-Liggett C, Knight R, Gordon JI (2007) The human microbiome project: exploring the microbial part of ourselves in a changing world. Nature 449 (7164):804–810
6. Qin J et al (2010) A human gut microbial gene catalogue established by metagenomic sequencing. Nature 464(7285):59–65

7. Rieder R, Wisniewski PJ, Alderman BL, Campbell SC (2017) Microbes and mental health: a review. Brain Behav Immun, In Press
8. Dzutsev A, Badger JH, Perez-Chanona E et al (2017) Microbes and Cancer. Annu Rev Immunol 35:199–228
9. Tsilimigras MCB, Fodor AA (2016) Compositional data analysis of the microbiome: fundamentals, tools, and challenges. Ann Epidemiol 26(5):330–335
10. 2015 Microbiome, metagenomics, and high-dimensional compositional data analysis. Ann Rev Stat Appl 2(1):73–94
11. Xiao K-Q et al (2016) Metagenomic profiles of antibiotic resistance genes in paddy soils from South China. FEMS Microbiol Ecol 92(3), fiw023
12. Hamady M, Knight R (2009) Microbial community profiling for human microbiome projects: tools, techniques, and challenges. Genome Res 19(7):1141–1152
13. Machine learning techniques accurately classify microbial communities by bacterial vaginosis characteristics. [Online]. Available: http://journals.plos.org/plosone/article?id=10.1371/journal.pone.0087830. Accessed 04 Feb 2017
14. Jiang X, Hu X, Xu W, He T, Park EK (2013) Comparison of dimensional reduction methods for detecting and visualizing novel patterns in human and marine microbiome. IEEE Trans Nanobioscience 12(3):199–205
15. Tyler AD, Smith MI, Silverberg MS (2014) Analyzing the human microbiome: a 'How To' guide for physicians. Am J Gastroenterol 109(7):983–993
16. Bartram AK et al (2014) Exploring links between pH and bacterial community composition in soils from the Craibstone experimental farm. FEMS Microbiol Ecol 87(2):403–415
17. Jiang X et al (2012) Functional biogeography of ocean microbes revealed through non-negative matrix factorization. PLOS ONE 7(9):e43866
18. Jiang X, Weitz JS, Dushoff J (Mar. 2012) A non-negative matrix factorization framework for identifying modular patterns in metagenomic profile data. J Math Biol 64(4):697–711
19. Arumugam M et al (2011) Enterotypes of the human gut microbiome. Nature 473 (7346):174–180
20. Personalized microbial network inference via multi-view clustering of oral metagenomics data – TiFN. [Online]. Available: http://www.tifn.nl/publication/personalized-microbial-network-inference-via-multi-view-clustering-of-oral-metagenomics-data/. Accessed 04 Feb 2017
21. Kuang D, Ding C, Park H (2012) Symmetric nonnegative matrix factorization for graph clustering. In: Proceedings of the 2012 SIAM international conference on data mining (0 vols). Society for Industrial and Applied Mathematics. pp 106–117
22. Raes J, Letunic I, Yamada T, Jensen LJ, Bork P (2011) Toward molecular trait-based ecology through integration of biogeochemical, geographical and metagenomic data. Mol Syst Biol 7 (1):n/a–n/a
23. Patel PV, Gianoulis TA, Bjornson RD, Yip KY, Engelman DM, Gerstein MB (2010) Analysis of membrane proteins in metagenomics: networks of correlated environmental features and protein families. Genome Res 20(7):960–971
24. He X, Cai D, Yan S, Zhang H-J (2005) Neighborhood preserving embedding. In: Tenth IEEE International Conference on Computer Vision (ICCV'05) Volume 1 2:1208–1213. Vol. 2
25. Chen X, Hu X, Shen X, Rosen G (2010) Probabilistic topic modeling for genomic data interpretation. In: 2010 I.E. International Conference on Bioinformatics and Biomedicine, BIBM 2010, Hong Kong, China, December 18–21, 2010, Proceedings, pp 149–152
26. Temporal probabilistic modeling of bacterial compositions derived from 16S rRNA sequencing | bioRxiv. [Online]. Available: http://biorxiv.org/content/early/2016/09/22/076836. Accessed 04 Feb 2017
27. Dietert RR, Silbergeld EK (2015) Biomarkers for the 21st century: listening to the microbiome. Toxicol Sci 144(2):208–216
28. Jiang X, Hu X, Xu W, Wang Y (2013) Manifold-constrained regularization for variable selection in environmental microbiomic data. In: 2013 I.E. International Conference on Bioinformatics and Biomedicine, Shanghai, China, December 18–21, 2013, pp 86–89

29. Lin W, Shi P, Feng R, Li H (2014) Variable selection in regression with compositional covariates. Biometrika 101(4):785–797
30. Shi P, Zhang A, Li H (2016) Regression analysis for microbiome compositional data. arXiv:1603.00974 [stat]
31. Randolph TW, Zhao S, Copeland W, Hullar M, Shojaie A (2015) Kernel-penalized regression for analysis of microbiome data. arXiv:1511.00297 [stat]
32. Faust K, Raes J (2012) Microbial interactions: from networks to models. Nat Rev Micro 10 (8):538–550
33. Fuhrman JA (2009) Microbial community structure and its functional implications. Nature 459 (7244):193–199
34. Fritz JV, Desai MS, Shah P, Schneider JG, Wilmes P (2013) From meta-omics to causality: experimental models for human microbiome research. Microbiome 1:14
35. @MInter: automated text-mining of microbial interactions | Bioinformatics | Oxford Academic. [Online]. Available: https://academic.oup.com/bioinformatics/article-abstract/32/19/2981/2196520/MInter-automated-text-mining-of-microbial?redirectedFrom=fulltext. Accessed 04 Feb 2017
36. Cordero OX, Datta MS (2016) Microbial interactions and community assembly at microscales. Curr Opin Microbiol 31:227–234
37. NetCooperate: a network-based tool for inferring host-microbe and microbe-microbe cooperation | BMC Bioinformatics | Full Text. [Online]. Available: https://bmcbioinformatics.biomedcentral.com/articles/10.1186/s12859-015-0588-y. Accessed 04 Feb 2017
38. Orth JD, Thiele I, Palsson BØ (2010) What is flux balance analysis? Nat Biotechnol 28 (3):245–248
39. Constructing and analyzing metabolic flux models of microbial communities | KBase
40. Shoaie S, Nielsen J (2014) Elucidating the interactions between the human gut microbiota and its host through metabolic modeling. Front Genet 5
41. Gerber GK (2014) The dynamic microbiome. FEBS Lett 588(22):4131–4139
42. Incomplete recovery and individualized responses of the human distal gut microbiota to repeated antibiotic perturbation. [Online]. Available: http://www.pnas.org/content/108/Supplement_1/4554.short. Accessed 04 Feb 2017
43. Extended local similarity analysis (eLSA) of microbial community and other time series data with replicates | BMC Systems Biology | Full Text." [Online]. Available: https://bmcsystbiol.biomedcentral.com/articles/10.1186/1752-0509-5-S2-S15. Accessed 2017
44. Jiang X, Hu X, Xu W, Park EK (2015) Predicting microbial interactions using vector autoregressive model with graph regularization. IEEE/ACM Trans Comput Biology Bioinform 12(2):254–261
45. Ma Y, Hu X, He T et al (2016) Hessian regularization based symmetric nonnegative matrix factorization for clustering gene expression and microbiome data[J]. Methods 111:80–84
46. Rangel C et al (2004) Modeling T-cell activation using gene expression profiling and state-space models. Bioinformatics 20(9):1361–1372
47. Probabilistic Boolean networks: a rule-based uncertainty model for gene regulatory networks | Bioinformatics | Oxford Academic. [Online]. Available: https://academic.oup.com/bioinformatics/article/18/2/261/225574/Probabilistic-Boolean-networks-a-rule-based. Accessed 04 Feb 2017

# Chapter 6
# Ontology-Based Vaccine Adverse Event Representation and Analysis

Jiangan Xie and Yongqun He

**Abstract** Vaccine is the one of the greatest inventions of modern medicine that has contributed most to the relief of human misery and the exciting increase in life expectancy. In 1796, an English country physician, Edward Jenner, discovered that inoculating mankind with cowpox can protect them from smallpox (Riedel S, Edward Jenner and the history of smallpox and vaccination. Proceedings (Baylor University. Medical Center) 18(1):21, 2005). Based on the vaccination worldwide, we finally succeeded in the eradication of smallpox in 1977 (Henderson, Vaccine 29:D7–D9, 2011). Other disabling and lethal diseases, like poliomyelitis and measles, are targeted for eradication (Bonanni, Vaccine 17:S120–S125, 1999).

Although vaccine development and administration are tremendously successful and cost-effective practices to human health, no vaccine is 100% safe for everyone because each person reacts to vaccinations differently given different genetic background and health conditions. Although all licensed vaccines are generally safe for the majority of people, vaccinees may still suffer adverse events (AEs) in reaction to various vaccines, some of which can be serious or even fatal (Haber et al., Drug Saf 32(4):309–323, 2009). Hence, the double-edged sword of vaccination remains a concern.

To support integrative AE data collection and analysis, it is critical to adopt an AE normalization strategy. In the past decades, different controlled terminologies, including the Medical Dictionary for Regulatory Activities (MedDRA) (Brown EG, Wood L, Wood S, et al., Drug Saf 20(2):109–117, 1999), the Common Terminology Criteria for Adverse Events (CTCAE) (NCI, The Common Terminology Criteria for Adverse Events (CTCAE). Available from: http://evs.nci.nih.gov/ftp1/CTCAE/About.html. Access on 7 Oct 2015), and the World Health

---

J. Xie
School of Bioinformatics, Chongqing University of Posts and Telecommunications, Chongqing, China

University of Michigan Medical School, 1301 Medical School Research Building III, 1150 W, Medical Center Dr, Ann Arbor, MI 48109, USA

Y. He (✉)
University of Michigan Medical School, 1301 Medical School Research Building III, 1150 W, Medical Center Dr, Ann Arbor, MI 48109, USA
e-mail: yongqunh@med.umich.edu

Organization (WHO) Adverse Reactions Terminology (WHO-ART) (WHO, The WHO Adverse Reaction Terminology – WHO-ART. Available from: https://www.umc-products.com/graphics/28010.pdf), have been developed with a specific aim to standardize AE categorization. However, these controlled terminologies have many drawbacks, such as lack of textual definitions, poorly defined hierarchies, and lack of semantic axioms that provide logical relations among terms. A biomedical ontology is a set of consensus-based and computer and human interpretable terms and relations that represent entities in a specific biomedical domain and how they relate each other. To represent and analyze vaccine adverse events (VAEs), our research group has initiated and led the development of a community-based ontology: the Ontology of Adverse Events (OAE) (He et al., J Biomed Semant 5:29, 2014). The OAE has been found to have advantages to overcome the drawbacks of those controlled terminologies (He et al., Curr Pharmacol Rep :1–16. doi:10.1007/s40495-016-0055-0, 2014) . By expanding the OAE and the community-based Vaccine Ontology (VO) (He et al., VO: vaccine ontology. In The 1st International Conference on Biomedical Ontology (ICBO-2009). Nature Precedings, Buffalo. http://precedings.nature.com/documents/3552/version/1; J Biomed Semant 2 (Suppl 2):S8; J Biomed Semant 3(1):17, 2009; Ozgur et al., J Biomed Semant 2 (2):S8, 2011; Lin Y, He Y, J Biomed Semant 3(1):17, 2012), we have also developed the Ontology of Vaccine Adverse Events (OVAE) to represent known VAEs associated with licensed vaccines (Marcos E, Zhao B, He Y, J Biomed Semant 4:40, 2013).

In this book chapter, we will first introduce the basic information of VAEs, VAE safety surveillance systems, and how to specifically query and analyze VAEs using the US VAE database VAERS (Chen et al., Vaccine 12(10):960–960, 1994). In the second half of the chapter, we will introduce the development and applications of the OAE and OVAE. Throughout this chapter, we will use the influenza vaccine Flublok as the vaccine example to launch the corresponding elaboration (Huber VC, McCullers JA, Curr Opin Mol Ther 10(1):75–85, 2008). Flublok is a recombinant hemagglutinin influenza vaccine indicated for active immunization against disease caused by influenza virus subtypes A and type B. On January 16, 2013, Flublok was approved by the FDA for the prevention of seasonal influenza in people 18 years and older in the USA. Now, more than 3 years later, an exploration of the reported AEs associated with this vaccine is urgently needed.

**Keywords** Vaccine Administration • Ontology • Adverse event • Flublok

## 6.1 Introduction of VAEs and VAE Case Report Resources

### 6.1.1 A VAE Does Not Have to Be Causal

According to the definition of AE from the Food and Drug Administration (FDA) (http://www.fda.gov/Safety/MedWatch/HowToReport/ucm053087.htm), an AE is any undesirable experience associated with the use of a medical product in a patient, and a serious adverse event (SAE) is an AE that results in serious or fatal health condition such as death, life-threatening, permanent damage, congenital anomaly, or hospitalization. In this chapter, a vaccine AE (VAE) refers to an AE associated with a vaccine and does not have to have a causal relation with an administration of the vaccine. Understanding the VAE profiles associated with special vaccines is crucial to predict potential VAEs, especially SAEs [4], and to improve vaccine safety.

## 6.1.2 VAE Data Resources

As a society, we need to know how many people become sick after vaccination with special attention on death and permanent injury. In addition, we would also like to know what AEs could occur and the probabilities of occurrence of these AEs. In the USA, in order to monitor post-marketing AEs associated with licensed vaccines, the National Vaccine Information Center (NVIC) worked with the US Congress to pass the National Childhood Vaccine Injury Act of 1986, which urges the creation of safety provisions for childhood vaccinations. Soon after, the Vaccine Adverse Event Reporting System (VAERS), a centralized vaccine reaction reporting system, was jointly established by the federal FDA and Centers for Disease Control and Prevention (CDC) in 1990 [14]. In recent years, VAERS receives over 30,000 AE case reports annually from healthcare providers, vaccine recipients, vaccine manufacturers, and other interested parties. To date, greater than 9000 AE symptoms associated with more than 80 vaccine products have been recorded in the VAERS database.

In addition to VAERS, many other VAE surveillance systems and databases have been developed and used by different organizations or countries. For example, the Canadian Adverse Events Following Immunization Surveillance System (CAEFISS; http://www.phac-aspc.gc.ca/im/vs-sv/index-eng.php) accepts VAE case reports in Canada. The UK Medicines and Healthcare products Regulatory Agency (MHRA; http://www.gov.uk/mhra) is responsible for accepting and monitoring VAEs that happen in the UK. It is noted that existing VAE case report systems are all passive surveillance systems that are subject to a number of well-described limitations, including variability in report quality, biased reporting, underreporting, and the inability in VAE causality determination [16]. Therefore, it is usually impossible to determine causal associations between vaccines and AEs using the data in these databases. However, statistical data analyses using the data from these databases are able to provide important clues and rationale for further systematical studies and investigations.

## 6.2 VAERS Data Report and Query

### 6.2.1 VAE Reporting to VAERS

It is very important that all serious health problems identified after vaccination in the USA be reported to VAERS. The US law requires doctors and other vaccine providers to report hospitalizations, injuries, deaths, and serious health problems following vaccination to VAERS (http://www.nvic.org/reportreaction.aspx). A VAE case report should be submitted if a child has regressed physically, mentally, or emotionally after vaccination or has experienced acute symptoms within hours, days or weeks of vaccination. The VAE case reporting to VAERS can be done by three ways: online submission, fax, and mail (https://vaers.hhs.gov/esub/index).

### 6.2.2 VAERS Data Search

The data stored in the VAERS database is available for anyone to search or download. The government's CDC Wonder search engine (http://wonder.cdc.gov/vaers.html) is the most commonly used VAERS database search engine. Two VAERS data search methods are provided: (i) VAERS Data Search, and (ii) VAERS Report Details search.

The VAERS Data Search (http://wonder.cdc.gov/controller/datarequest/D8) can be searched by the following: age, event category, gender, manufacturers, onset interval, recovery status, serious/nonserious category, state/territory, symptoms, vaccine, year reported, month reported, year vaccinated, and month vaccinated. The request form includes 12 input sections, each of which has several different parameter settings. Take the influenza vaccine Flublok [15] as an example (Fig. 6.1), for which we provided specific information for the following criteria: (i) vaccine, Flublok (Fig. 6.1a); (ii) age, over 18 years old (Fig. 6.1b); and (iii) report received date, January 2013 to February 2016 (Fig. 6.1c). The default values are taken for the other input sections. Our search identified that 77 AE cases consisting of 129 AE symptoms were reported to VAERS during the queried time period. Figure 6.1d provides a screenshot showing partial results of such a query. More analysis on this set of data will be described in the following sections.

The VAERS Report Details search (http://wonder.cdc.gov/controller/datarequest/D8) allows the search of the details of a specific VAERS case report based on a VAERS ID number. Figure 6.2 shows the search process and results given a VAERS ID: 123456-1. The results of such a query include all the details, such as the patient age, vaccine type, vaccination data, and event categories (Fig. 6.2).

A more detailed tutorial of CDC's VAERS Wonder system is available at the website: https://vaers.hhs.gov/data/VAERS_WONDER_Text_Tutorial_2015.pdf.

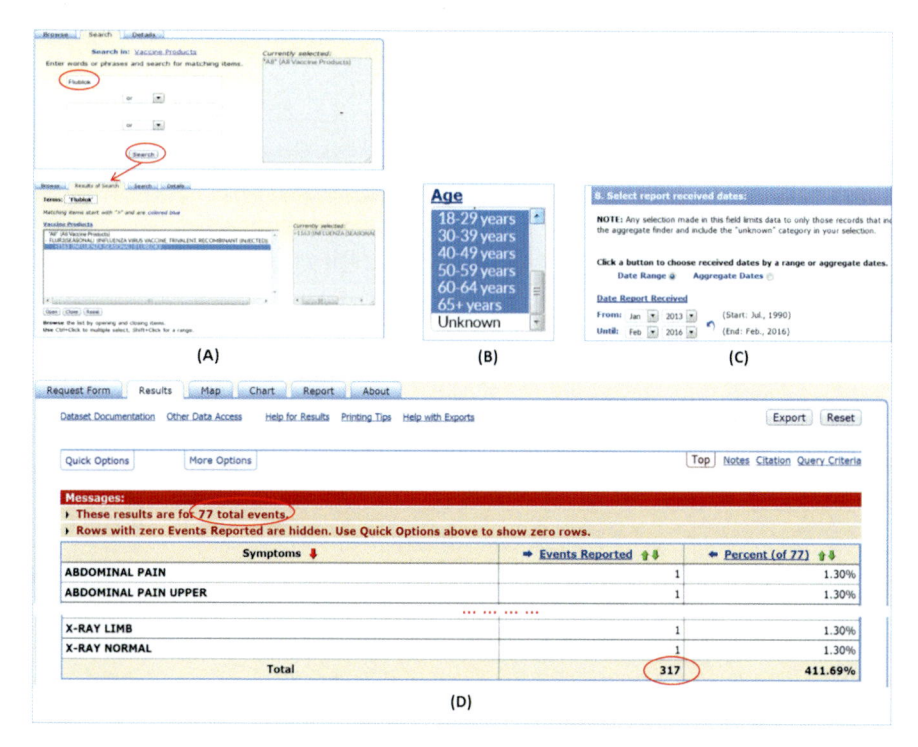

Fig. 6.1 Example of querying Flublok-associated AEs using VAERS Data Search from CDC Wonder. (a) The word "Flublok" was typed in the search program of the section of "3. Select vaccine characteristics." (b) The age of vaccination was set to be no less than 18 years old in "4. Select location, age, gender." (c) The date report received of AEs was set to be from January 2013 to February 2016 in "8. Select report received dates." The default values were used for all other query criteria. (d) Search results based on above setting conditions were provided. This query was conducted on April 16, 2016, on the CDC Wonder website: http://wonder.cdc.gov/controller/datarequest/D8

Furthermore, the US National Vaccine Information Center (NVIC), a national charitable, nonprofit educational organization founded in 1982 in the USA, hosts another VAERS database search engine: MedAlerts (http://www.medalerts.org/vaersdb/index.php). Compared to the CDC Wonder VAERS Data Search program, MedAlerts is simpler and user-friendly and offers powerful search and reporting capabilities.

## 6.3 VAE Data Statistical Analysis

Since most reported VAE data were obtained by clinical observations from various people such as physicians and nonprofessional vaccinees or vaccinees' parents, these data may include many biased reporting and noises. Direct and naïve usages

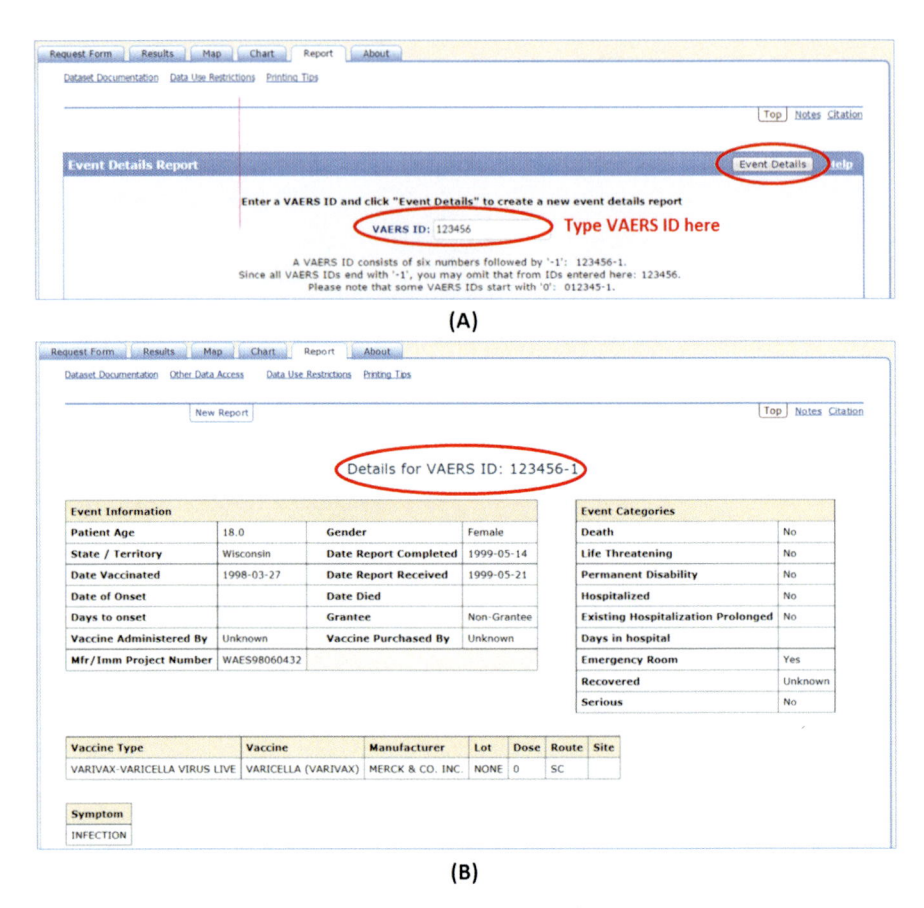

**Fig. 6.2** Example of querying detailed information of individual VAE case report using VAERS Report Details program from CDC Wonder. (**a**) A VAERS ID (123456-1) was typed in the search box, followed by a click on the Event Details. (**b**) Detailed information for the VAERS ID was provided. This query was conducted on April 16, 2016, on the CDC Wonder website: http://wonder.cdc.gov/controller/datarequest/D8

of these temporally associated data may result in wrong assertion of causal relations between AEs and vaccines. Fortunately, by combining multiple statistical and bioinformatics methods, background noise and irrelevant information in these data can be reduced to a minimal level. Statistically processed and analyzed data are more meaningful and interpretable and can be used to draw sensible hypotheses. This section introduces commonly used VAE data analysis methods.

## 6.3.1 Minimal Case Report Number Filtration

In VAE data analysis, the minimal case report number (or base level) filtration method is essential to filter out background noises. A commonly used cutoff is three case reports for each AE to be further considered. However, such a constant cutoff does not work well for the total number of case reports that has a big size. When the total case report number is greater than 1500, the cutoff is often set to be 0.2% of total reports. Such a cutoff means that at least 2 out of 1000 cases should report the AE of interest [17, 18].

In the Flublok example, since VAERS included only 77 AE case reports (Fig. 6.1d), we used three case reports as the cutoff for selecting AEs for further analysis.

## 6.3.2 Proportional Reporting Ratio (PRR)

The proportional reporting ratio (PRR) approach was finalized by Evans et al. [19] and has now become a popular method for signal generation from spontaneous adverse drug reaction reports. The combined use of PRR and a minimal sample size cutoff is also called the screened PRR method (SPRR) [20].

The VAE data in VAERS database can be viewed as a contingency table with rows representing the AE coding terms and columns representing the vaccine products. Each cell in the table contains a value that gives the number of reports for the AE of interest and the vaccine of interest. Using a $2 \times 2$ contingency table data structure (Table 6.1), PRR calculates the proportion of a specific AE for a vaccine of interest, where the comparator is all other vaccines in the VAERS database [17, 20, 21]:

$$PRR = \frac{a/(a+c)}{b/(b+d)} \quad (6.1)$$

A large PRR score of a specific vaccine AE indicates that the AE has been disproportionately reported for that vaccine, compared with all of the other vaccines in the VAERS database.

For example, suppose that diarrhea was reported 100 times for a given vaccine of interest, out of 2000 AEs reported for the vaccine. Thus the proportion of diarrhea AE for this vaccine is $100/2000 = 0.05$. Suppose that in a VAE database (e.g., VAERS) diarrhea was reported 1000 times, out of 100,000 total AE case reports for all vaccines in the database. Thus, diarrhea was reported with proportion $1000/100000 = 0.01$ for all vaccines in the database. The PRR in this case is $0.05/0.01 = 5$. This tells us that diarrhea was reported five times as frequently (among all AE reports) for the vaccine of interest compared to the vaccines in the comparison group.

**Table 6.1** Calculation of PRR and $x^2$ for vaccine adverse events

|  | Vaccine of interest | All other vaccines in VAERS |
|---|---|---|
| Adverse event of interest | a | b |
| All other adverse events | c | d |

### 6.3.3 Chi-Square ($x^2$) Test

In parallel with PRR signal detection, a $x^2$ test was applied to statistically analyze the likelihood of individual AE terms associated with specific vaccines [17, 20]. The $x^2$ test was introduced by Pearson in 1900 [22], with the vital modification as to degrees of freedom established by Fisher in 1922 [23]. The $x^2$ test following the Yates correction, which is a continuity adjustment to improve the accuracy of the Chi-square approximation, was later introduced into the distribution of the Pearson's test [24].

Based on the same $2 \times 2$ contingency table used in the PRR calculation (Table 6.1), the $x^2$ test is calculated by the following formula:

$$x^2 = \frac{(ad - bc)^2 (a + b + c + d)}{(a + b)(c + d)(b + d)(a + c)} \tag{6.2}$$

An AE is called significant when its $x^2$ score is greater than 4, which is equivalent to a $p$-value of approximately 0.05 or smaller [17].

For the Flublok example, by adopting the screening criteria of minimal case report number of 3, the PRR score $>= 2$, and the $x^2 > = 4$, we identified 18 statistically significant AEs associated with the Flublok (Table 6.2).

## 6.4 MedDRA and OAE for Vaccine AE Standardization

MedDRA is the most widely used medical terminology in annotating AE information [5–7]. MedDRA has played a central role in standardizing and improving vocabulary in the scope of VAE reporting and recording. In VAERS database, all signs and symptoms of submitted AE reports were coded using MedDRA. VAERS so far has used 9893 MedDRA terms for AE standardization.

The Ontology of Adverse Events (OAE) is a community-driven ontology developed to standardize and integrate data relating to AEs arising subsequent to medical interventions as well as to support computer-assisted reasoning [8]. In OAE, an AE is represented as a pathological bodily process in a patient that occurs after a medical intervention (e.g., vaccination) and has an outcome including symptom (e.g., vomiting), sign (e.g., increased heart rate), or an abnormal process (e.g., viral infection). The OAE-defined AE does not have to be causally induced by a medical

**Table 6.2** Flublok-specific adverse events

| Adverse event | Count | PRR | $x^2$ |
|---|---|---|---|
| Anaphylactic reaction | 4 | 22.139 | 78.382 |
| Burning sensation | 3 | 3.03 | 4.115 |
| Dysphonia | 4 | 9.902 | 31.722 |
| Dyspnea | 7 | 2.623 | 7.253 |
| Flushing | 4 | 6.483 | 18.526 |
| Inflammation | 3 | 5.446 | 10.881 |
| Lip swelling | 4 | 7.481 | 22.371 |
| Local swelling | 4 | 4.647 | 11.5 |
| Paresthesia oral | 3 | 9.633 | 22.982 |
| Pharyngeal edema | 4 | 6.774 | 19.644 |
| Pruritus | 13 | 2.679 | 14.549 |
| Rash | 14 | 2.771 | 16.888 |
| Rash erythematous | 5 | 2.846 | 6.102 |
| Rash macular | 3 | 6.208 | 13.07 |
| Rash pruritic | 4 | 3.846 | 8.492 |
| Swelling face | 4 | 5.435 | 14.502 |
| Throat tightness | 3 | 4.695 | 8.737 |
| Urticaria | 10 | 3.181 | 15.521 |

intervention, which is consistent with its definition in commonly used clinical scenarios. As of April 10, 2016, OAE has 5637 terms.

Compared to MedDRA, OAE offers better hierarchical classification of AEs [9, 17, 25, 26]. MedDRA has several issues related to the usage of MedDRA in domain completeness and discrepancies with a physician's AE description [27]. Firstly, MedDRA does not provide definitions for its coding terms, which may cause confusion when the terms are assigned to the VAERS cases. Secondly, the poorly defined hierarchical structures of MedDRA make it difficult to use for advanced analysis such as AE classification. A previous study empirically compared the MedDRA and OAE in AE classification [17]. This study analyzed VAEs associated with the administration of two different types of seasonal influenza vaccines: trivalent (killed) inactivated influenza vaccine (TIV) and trivalent live attenuated influenza vaccine (LAIV). The statistical analysis identified 48 - TIV-enriched and 68 LAIV-enriched AEs using the data in the VAERS database. After the mapping of the MedDRA terms related to these statistically significant AEs to OAE terms, the hierarchy structure of these terms was generated based on OAE and compared with the MedDRA-derived hierarchy. The results demonstrate the advantages of OAE compared to MedDRA in AE classification [17].

Figure 6.3a represents the OAE-based hierarchical structure of 18 statistically significant AEs associated with Flublok. The most frequent AE diagnostic category was the skin system (e.g., pruritus, erythematous rash, urticaria, and flushing), followed by the homeostasis system (e.g., lip swelling and swelling face). All of these AEs are mild and self-limited. No serious AEs following Flublok vaccination have been identified as statistically significant.

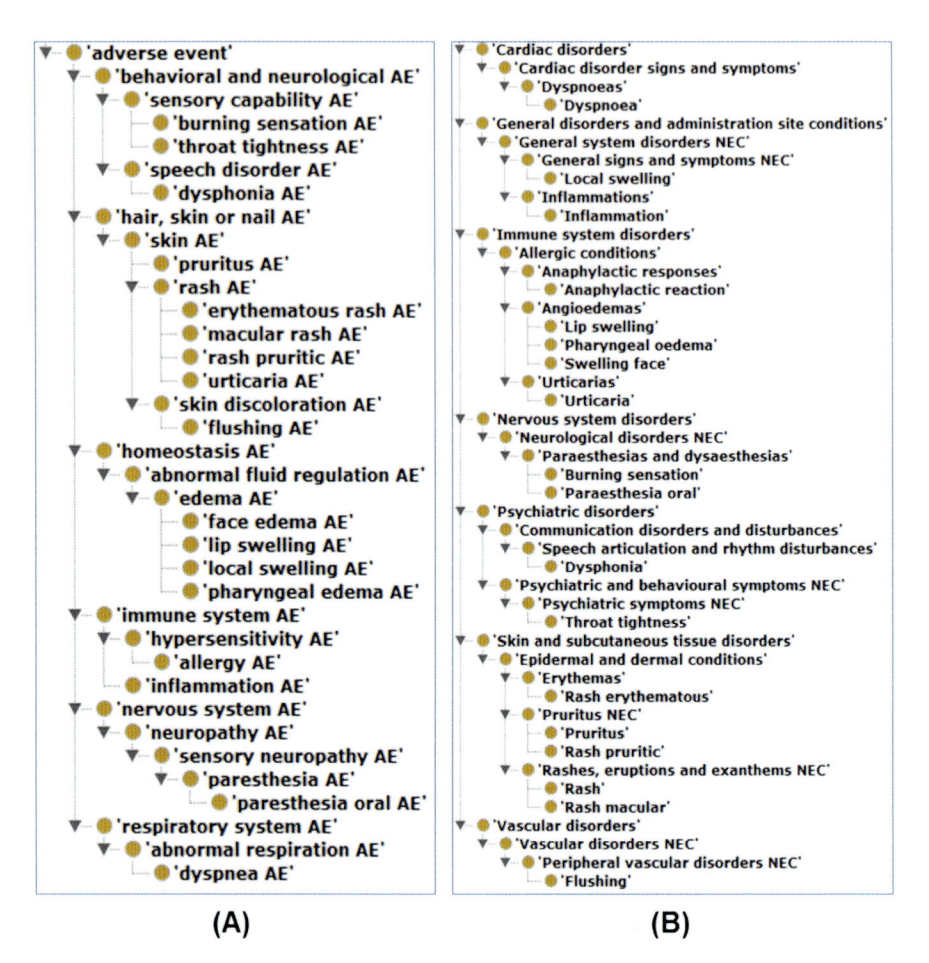

```
▼ ● 'adverse event'
  ▼ ● 'behavioral and neurological AE'
    ▼ ● 'sensory capability AE'
      ● 'burning sensation AE'
      ● 'throat tightness AE'
    ▼ ● 'speech disorder AE'
      ● 'dysphonia AE'
  ▼ ● 'hair, skin or nail AE'
    ▼ ● 'skin AE'
      ● 'pruritus AE'
      ▼ ● 'rash AE'
        ● 'erythematous rash AE'
        ● 'macular rash AE'
        ● 'rash pruritic AE'
        ● 'urticaria AE'
      ▼ ● 'skin discoloration AE'
        ● 'flushing AE'
  ▼ ● 'homeostasis AE'
    ▼ ● 'abnormal fluid regulation AE'
      ▼ ● 'edema AE'
        ● 'face edema AE'
        ● 'lip swelling AE'
        ● 'local swelling AE'
        ● 'pharyngeal edema AE'
  ▼ ● 'immune system AE'
    ▼ ● 'hypersensitivity AE'
      ● 'allergy AE'
    ● 'inflammation AE'
  ▼ ● 'nervous system AE'
    ▼ ● 'neuropathy AE'
      ▼ ● 'sensory neuropathy AE'
        ▼ ● 'paresthesia AE'
          ● 'paresthesia oral AE'
  ▼ ● 'respiratory system AE'
    ▼ ● 'abnormal respiration AE'
      ● 'dyspnea AE'
```

(A)

```
▼ ● 'Cardiac disorders'
  ▼ ● 'Cardiac disorder signs and symptoms'
    ▼ ● 'Dyspnoeas'
      ● 'Dyspnoea'
▼ ● 'General disorders and administration site conditions'
  ▼ ● 'General system disorders NEC'
    ▼ ● 'General signs and symptoms NEC'
      ● 'Local swelling'
    ▼ ● 'Inflammations'
      ● 'Inflammation'
▼ ● 'Immune system disorders'
  ▼ ● 'Allergic conditions'
    ▼ ● 'Anaphylactic responses'
      ● 'Anaphylactic reaction'
    ▼ ● 'Angioedemas'
      ● 'Lip swelling'
      ● 'Pharyngeal oedema'
      ● 'Swelling face'
    ▼ ● 'Urticarias'
      ● 'Urticaria'
▼ ● 'Nervous system disorders'
  ▼ ● 'Neurological disorders NEC'
    ▼ ● 'Paraesthesias and dysaesthesias'
      ● 'Burning sensation'
      ● 'Paraesthesia oral'
▼ ● 'Psychiatric disorders'
  ▼ ● 'Communication disorders and disturbances'
    ▼ ● 'Speech articulation and rhythm disturbances'
      ● 'Dysphonia'
  ▼ ● 'Psychiatric and behavioural symptoms NEC'
    ▼ ● 'Psychiatric symptoms NEC'
      ● 'Throat tightness'
▼ ● 'Skin and subcutaneous tissue disorders'
  ▼ ● 'Epidermal and dermal conditions'
    ▼ ● 'Erythemas'
      ● 'Rash erythematous'
    ▼ ● 'Pruritus NEC'
      ● 'Pruritus'
      ● 'Rash pruritic'
    ▼ ● 'Rashes, eruptions and exanthems NEC'
      ● 'Rash'
      ● 'Rash macular'
▼ ● 'Vascular disorders'
  ▼ ● 'Vascular disorders NEC'
    ▼ ● 'Peripheral vascular disorders NEC'
      ● 'Flushing'
```

(B)

**Fig. 6.3** Hierarchical classification of Flublok-specific AEs based on OAE and MedDRA. (**a**) Asserted OAE hierarchical structure. Note that OAE also includes axioms and axiom-based reasoning features. Such features are not described here. Interested readers can find more information from the OAE paper [8]. (**b**) MedDRA hierarchical structure. OntoFox (http://ontofox.hegroup.org/) [33] was used to extract a subset of OAE that includes Flublok-specific AEs, related top-level OAE terms, and their relations. The Protégé OWL editor (http://protege.stanford.edu/) was used to display the hierarchies and generate the screenshots

The Flublok example is used here to further illustrate the difference between MedDRA and OAE in AE classification, as shown in the MedDRA classification of the 18 Flublok-specific AEs (Fig. 6.3b). MedDRA includes many terms ended with "NEC" (i.e., "not elsewhere classified"), for example, "general system disorders NEC" and "neurological disorders NEC." Such an "NEC" term definition style is arbitrary and ambiguous and often leads to confusing and unclear classification results. For example, the parent MedDRA term of "pruritus" is "pruritus NEC," which is confusing and logically incorrect. MedDRA also contains many redundant terms and hierarchical structures, such as "urticaria" that is listed as a subclass of "urticarias" (Fig. 6.3b); this is a meaningless repetition. To represent a type of entity, a class term is typically represented as a singular noun instead of a plural noun in ontology. In addition, MedDRA misses obvious parent-child term logic. For example, MedDRA classifies "rash" and "rash macular" in the same hierarchical level (Fig. 6.3b). This is logically incorrect since in reality, a "rash macular" is a subclass of "rash" (Fig. 6.3a). In comparison, such drawbacks are avoided and overcome in OAE.

## 6.5 OVAE and Its Application in VAE Data Analysis

Many known VAEs at the population level have been recorded in the package insert documents of commercial vaccine products. Compared to the noisy data from clinical VAE case reports, the VAEs recorded in the official package insert documents are typically generated from randomized and well-controlled clinical trials. Therefore, these VAEs are considered as reliable known AEs specific for individual vaccines.

The Ontology of Vaccine Adverse Events (OVAE) is developed to reuse many terms from the OAE and the VO [10–12], and it logically represents the AEs recorded in the official package inserts of US-licensed human vaccines [13]. At present, OVAE includes over 1300 AEs associated with 63 US-licensed human vaccines [13]. Figure 6.4 provides an example of representing Flublok-associated fatigue AE. The term "Flublok-associated fatigue AE" (Fig. 6.4a) is an OAE-specific term that has the ontology term unique resource identifier (URI) http://purl.obolibrary.org/obo/OVAE_0000240. A click of this URI links to a website in Ontobee [28]: http://www.ontobee.org/ontology/OVAE?iri=http://purl.obolibrary.org/obo/OVA E_0000240. In this case, the Ontobee server is called the default linked data server for OVAE. Figure 6.4b demonstrates three ontological axioms for the OAE term "Flublok-associated fatigue AE." An ontological axiom is a statement that provides explicit logical assertion using ontology terms and properties (also called relations). The axioms shown in Fig. 6.4b define "Flublok-associated fatigue AE" as the subclasses of "fatigue AE" and "Flublok vaccine adverse event," and they also define different occurrences of the Flublok at the populations with different age ranges, for example, the VAE occurrence of 0.15 (or 15%) at the age of 18–49 years old. Note that the occurrence information is obtained by manual annotation of the AE reports shown in the Flublok package insert document (Fig. 6.4c) (http://www.fda.gov/downloads/BiologicsBlood Vaccines/Vaccines/ApprovedProducts/UCM336020.pdf).

The OVAE results may differ from statistically identified results using VAERS data. For the Flublok example, OVAE shows five AEs associated with this vaccine: headache, fatigue, muscle pain, injection-site pain, and myalgia. However, none of these five AEs are shown in the 18 AEs identified from our VAERS data analysis (Table 6.2). Such a comparative result indicates that VAERs turn to find more and

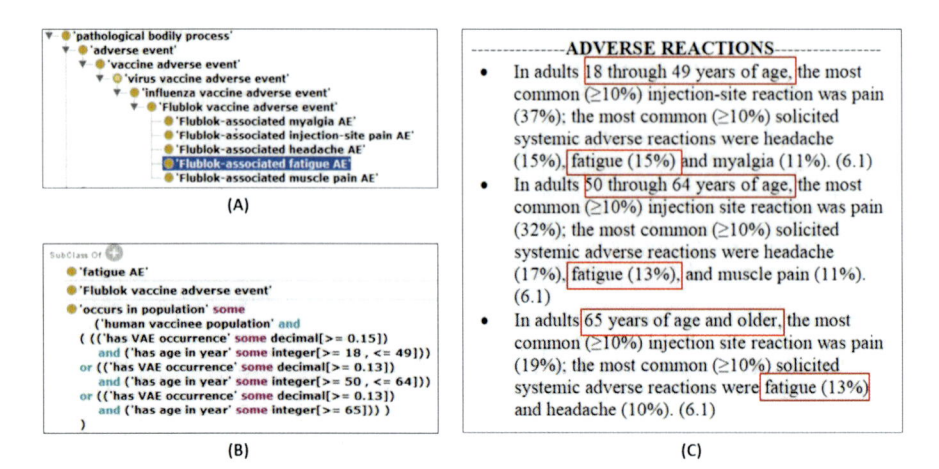

Fig. 6.4 OVAE representing Flublok VAEs recorded in FDA vaccine package insert. (a) The hierarchical structure of Flublok VAEs represented in OVAE. (b) OVAE axiom representation of "Flublok-associated fatigue AE" based on three age groups. (c) Flublok AEs recorded in the FDA package insert document. The highlighted age-dependent fatigue records in (c) are represented in OVAE as shown in (b). The subfigures (a) and (b) were screenshots of OVAE using the Protégé OWL editor. The text from (c) comes from the FDA package insert document of the Flublok vaccine

different AEs compared to the AEs reported to the FDA and described in the package insert documents.

## 6.6 Summary and Discussion

This chapter first introduces the background of VAEs, VAE data resources with a focus on VAERS, different statistical methods to analyze clinical VAE case report data, and how to use VAERS to query and analyze VAEs. We have used Flublok, a newly licensed influenza vaccine, as our use case study. In the past 3 years, only 77 case reports were recorded in VAERS. Such a low number of case reporting suggests that this vaccine is relatively safe. Flublok is a recombinant hemagglutinin influenza vaccine, but other existing licensed influenza vaccines are either killed whole organism vaccines or live attenuated vaccine. Compared to whole organism vaccines, recombinant protein vaccines are generally safe. Our analyzed results are consistent with such a general rule [29]. To better understand the differences of AE profiles associated with Flublok and other types of influenza vaccines, a thorough comparative study using the VAE case reports associated with different influenza vaccines is required.

This chapter also introduces existing AE-controlled terminologies (with a focus on MedDRA) and two VAE-related ontologies (OAE/OVAE). Our Flublok case study further illustrates the advantages of OAE over MedDRA in terms of AE classification. One drawback of OAE is its relatively low coverage compared to MedDRA. To make OAE fully used in the clinical AE case reporting and analysis, it is desired to extend OAE further to cover all possible AEs.

It is important to note that although different AEs are observed, overall vaccinations are very safe. The discovery of VAEs does not compromise the recommendation and requirement of vaccinations against various infectious diseases [1–3]. The identification of VAEs has at least two roles. First, the early identification of specific AEs, especially SAEs, would help the detection of potential issues such as poor vaccine quality and possible vaccine contaminations. Second, the detection of specific AEs may identify some severe AEs that are not detected in vaccine clinical trials. Although vaccine clinical trials are randomized, these trials usually recruit small populations of human subjects and may miss the sign of potentially severe AEs. Such severe AEs may be identified by using the large population case report data during a post-licensure vaccine safety surveillance stage.

One future challenge is to identify causal associations of some SAEs and vaccinations. Such a causal association is very likely due to the genetic variations and some specific health conditions in specific vaccinees. To better study AE causalities, the guidance of theories is critical. Two recently proposed theories, i.e., the "immune response gene network" theory [30, 31] and the OneNet Theory of Life [32], provide specific details on how to study molecular and biological VAE causal mechanisms. Ontologies and theories can be used in combination to form a semantic framework to integrate various clinical VAE data and genetic data under specific conditions [9]. The analysis of such data with ontology-based and theory-guided systems biology approaches will improve the mechanistic understanding of vaccine-specific AE profiles, further benefiting individual level VAE evaluation and prediction.

**Acknowledgment** We appreciate Mr. Jordan Rinder's editorial review and comments.

## References

1. Riedel S (2005) Edward Jenner and the history of smallpox and vaccination. Proceedings (Baylor University. Medical Center) 18(1):21
2. Henderson DA (2011) The eradication of smallpox – an overview of the past, present, and future. Vaccine 29:D7–D9
3. Bonanni P (1999) Demographic impact of vaccination: a review. Vaccine 17:S120–S125
4. Haber P et al (2009) Vaccines and Guillain-Barre syndrome. Drug Saf 32(4):309–323
5. Brown EG, Wood L, Wood S (1999) The medical dictionary for regulatory activities (MedDRA). Drug Saf 20(2):109–117
6. NCI. The Common Terminology Criteria for Adverse Events (CTCAE). Available from: http://evs.nci.nih.gov/ftp1/CTCAE/About.html. Access on 7 Oct 2015
7. WHO. The WHO Adverse Reaction Terminology – WHO-ART. Available from: https://www.umc-products.com/graphics/28010.pdf
8. He YQ et al (2014) OAE: the ontology of adverse events. J Biomed Semant 5:29

9. He Y (2016) Ontology-based vaccine and drug adverse event representation and theory-guided systematic causal network analysis toward integrative pharmacovigilance research. Curr Pharmacol Rep :1–16. doi:10.1007/s40495-016-0055-0). In press
10. He Y, et al (2009) VO: vaccine ontology. In The 1st International Conference on Biomedical Ontology (ICBO-2009). Nature Precedings, Buffalo. http://precedings.nature.com/documents/3552/version/1
11. Ozgur A et al (2011) Mining of vaccine-associated IFN-gamma gene interaction networks using the vaccine ontology. J Biomed Semant 2(2):S8
12. Lin Y, He Y (2012) Ontology representation and analysis of vaccine formulation and administration and their effects on vaccine immune responses. J Biomed Semant 3(1):17
13. Marcos E, Zhao B, He Y (2013) The Ontology of Vaccine Adverse Events (OVAE) and its usage in representing and analyzing adverse events associated with US-licensed human vaccines. J Biomed Semant 4:40
14. Chen RT et al (1994) The vaccine adverse event reporting system (Vaers) (Vol 12, Pg 542, 1994). Vaccine 12(10):960–960
15. Huber VC, McCullers JA (2008) FluBlok, a recombinant influenza vaccine. Curr Opin Mol Ther 10(1):75–85
16. Varricchio F et al (2004) Understanding vaccine safety information from the vaccine adverse event reporting system. Pediatr Infect Dis J 23(4):287–294
17. Sarntivijai S et al (2012) Ontology-based combinatorial comparative analysis of adverse events associated with killed and live influenza vaccines. PLoS One 7(11):e49941
18. Sarntivijai S et al (2016) Linking MedDRA®-coded clinical phenotypes to biological mechanisms by the ontology of adverse events: a pilot study on tyrosine kinase inhibitors. Drug Saf:1–11
19. Evans SJW, Waller PC, Davis S (2001) Use of proportional reporting ratios (PRRs) for signal generation from spontaneous adverse drug reaction reports. Pharmacoepidemiol Drug Saf 10(6):483–486
20. Banks D et al (2005) Comparing data mining methods on the VAERS database. Pharmacoepidemiol Drug Saf 14(9):601–609
21. Woo EJ et al (2008) Effects of stratification on data mining in the US Vaccine Adverse Event Reporting System (VAERS). Drug Saf 31(8):667–674
22. Pearson K (1900) X. on the criterion that a given system of deviations from the probable in the case of a correlated system of variables is such that it can be reasonably supposed to have arisen from random sampling. Lond, Edinb Dublin Philos Mag J Sci 50(302):157–175
23. Fisher RA (1922) On the interpretation of $\chi 2$ from contingency tables, and the calculation of P. J R Stat Soc 85(1):87–94
24. Yates F (1934) Contingency tables involving small numbers and the $\chi2$ test. Suppl J R Stat Soc 1:217–235
25. Xie J et al (2016) Differential adverse event profiles associated with BCG as a preventive tuberculosis vaccine or therapeutic bladder cancer vaccine identified by comparative ontology-based VAERS and literature meta-analysis. PLoS One 11(10):e0164792
26. Xie J et al (2016) Statistical and ontological analysis of adverse events associated with monovalent and combination vaccines against hepatitis A and B diseases. Sci Rep 6
27. Brown EG (2003) Methods and pitfalls in searching drug safety databases utilising the Medical Dictionary for Regulatory Activities (MedDRA). Drug Saf 26(3):145–158
28. Xiang Z, et al (2011) Ontobee: a linked data server and browser for ontology terms. In the 2nd International Conference on Biomedical Ontologies (ICBO). CEUR Workshop Proceedings, New York
29. Cox MM et al (2015) Safety, efficacy, and immunogenicity of Flublok in the prevention of seasonal influenza in adults. Ther Adv Vaccines 3(4):97–108
30. Poland GA et al (2007) Heterogeneity in vaccine immune response: the role of immunogenetics and the emerging field of vaccinomics. Clin Pharmacol Ther 82(6):653–664

31. Poland GA, Ovsyannikova IG, Jacobson RM (2009) Application of pharmacogenomics to vaccines. Pharmacogenomics 10(5):837–852
32. He Y (2014) Ontology-supported research on vaccine efficacy, safety and integrative biological networks. Expert Rev Vaccines 13(7):825–841
33. Xiang Z et al (2010) OntoFox: web-based support for ontology reuse. BMC Res Notes 3(1):175

# Chapter 7
# LEMRG: Decision Rule Generation Algorithm for Mining MicroRNA Expression Data

Łukasz Piątek and Jerzy W. Grzymała-Busse

**Abstract** Recently, research on mining *microRNA* (or *miRNA*) expression data has received a lot of attention, mainly because of its role in gene regulation. However, such type of data – usually saved in the form of *microarrays* – are very specific, because they contain only a small number of cases (often less than 100) compared with large number of attributes (equal to several hundreds or even tens of thousand). The small number of cases available during the learning process can cause instability of the newly created classifiers. Secondly, the huge number of attributes imposes the necessity of selecting only a few dominant attributes strongly correlated with the decision. Thus, an application of fundamental machine learning approaches of mining microarray data and its further classification is problematic or even could just fail.

Thus, the main goal of our research is to develop the generalized algorithm of mining microarray data (including *miRNA* data sets), mainly to improve stability and, consequently, accuracy of classification for the newly created learning classifiers. The main concept of the novel approach is based on iteratively inducing many subsequent decision rule sets – called decision *rule generations* – instead of inducing only a single decision rule set, as it is done routinely. The decision rules have been chosen as the baseline classifiers of the newly developed *LEMRG* (*Learning from Examples Module based on Rule Generations*) algorithm mainly because the decision rule-based knowledge representation is easier for humans to comprehend, rather than other learning models. In our research we used a *miRNA* expression level learning data set describing 11 types of human cancers, while the

Ł. Piątek (✉)
Institute für Biomedizinische Technik und Informatik, Technische Universität Ilmenau, Gustav-Kirchoff 2 Str, 98684 Ilmenau, Germany

Department of Expert Systems and Artificial Intelligence, University of Information Technology and Management, H. Sucharskiego 2 Str, 35225 Rzeszów, Poland
e-mail: lukasz.piatek@tu-ilmenau.de

J.W. Grzymała-Busse
Department of Electrical Engineering and Computer Science, University of Kansas, 3014 Eaton Hall, 1520 W. 15th Str, Lawrence, KS 66045-7621, USA

Department of Expert Systems and Artificial Intelligence, University of Information Technology and Management, H. Sucharskiego 2 Str, 35225 Rzeszów, Poland

testing data set contained poorly differentiated cases of only four types of cancers. As expected, our new classifiers – saved in the form of so-called *cumulative* decision rule sets – had better *stability* and *accuracy* of classification than single decision rule sets induced in the traditional manner. Furthermore, the *LEMRG* was compared with other machine learning models. It was proven that only 3 out of all 16 tested classifiers enabled so effective classification as our newly developed approach. Thus, using our *cumulative* set of decision rules, all cases of cancer from two selected concepts – *colon* and *ovary* – were correctly classified. Furthermore, we showed the role of these selected *miRNA*s as the potential biomarkers for diagnosis of tumors.

A preliminary result of our research on decision *rule generations* was initially presented at the first International Conference of *Digital Medicine and Medical 3D Printing* (17-19.06.2016, Nanjing, China).

**Keywords** MicroRNA • *MiRNA* • Decision *rule generations* • *Cumulative* decision rule sets • Data mining • Induction of decision rules • *LEM2* • *MLEM2* • *GTS* • *AQ* • *LEMRG*

## 7.1 Introduction

We begin this chapter by providing an overview of the current state of machine learning approaches used for *miRNA*s analysis. We would like to provide the reader with a brief background of the inherent challenges and potential of mining *miRNA* expression data. Afterward, we present the main idea and then the results of our research by an example of mining the *miRNA* data set describing human cancers [1]. In brief, the main idea of our novel approach to data mining is based on iteratively inducing many subsequent decision rule sets – called decision *rule generations* – instead of inducing only a single decision rule set in the regular way, i.e., as it is done routinely (explained in details in Subchapter 7.2 "Learning from Examples Based on Rule Generations"). Based on our preliminary results [2], it is expected that the newly developed classifiers, saved in the form of a *cumulative* set of decision rules, will be characterized by better stability of classification and then consequently higher classification accuracy (i.e., smaller classification error rate) than the single decision rule sets induced in the traditional manner. Thus, with respect to our problem and proposed solution to it, it is necessary to describe first issues related to:

- Limitations of classifiers in general, but especially those created for the *miRNA* data sets.
- Possible imperfections of data sets (*incompleteness*, *inconsistency*, and *numerical* attributes).
- Selected supervised machine learning models (*decision rules*).
- Results of our research on inducing decision *rule generations* for *miRNA* data sets.

We start by providing a few facts about *microarray* technology and *miRNA* data sets and some selected machine learning approaches used for analyzing and classifying such type of data.

### 7.1.1 Microarrays

The microarray technology allows to inspect gene expressions on a genomic level [3]. For instance, huge-scale profiling of human gene expressions allows to differentiate normal cells from cancer cells and inspect cellular changes in the progression of cancer metastasis [4]. The possibility of simultaneously inspecting a few thousands of genes is an indisputable advantage. However, further analysis of microarray data could be problematic, as it is a big challenge for the machine learning algorithms and computer expert systems. The very important problem of mining microarray data is the small number of cases compared with a huge number of features (attributes). Moreover, usually only a few selected genes are essential for the examined diseases. In the literature, algorithms used for gene selection or classification were divided into three main subgroups [5], including:

- *Filter* methods – uncovering dependencies by using statistical methods of ranking genes, instead of classification (e.g., class separability).
- *Wrapper* methods – classification based on using classifiers such as *Bayesian* classifier, *kNN* (*k-nearest neighbors*), or *SVM* (*support vector machine*).
- *Embedded* methods – gene selection performed by using *logistic regression* or *regularization*.

The filter feature selection methods allow to perform feature discovery in *GEM* (*gene expression microarray*) analysis, also called *DFG*s (*differentially expressed genes*) discovery, based on gene prioritization or biomarkers discovery [6]. However, the machine learning algorithms useful in analysis of multiarrays require often using some specific approaches. For example, higher efficiency in classification of gene expression data can be achieved by using (i) *aggregated* classifiers or (ii) *hybrid* classifiers. At first, rather than using a classifier in the form of only single *DT* (*decision tree*), Huang et al. used decision forest [7] and Stiglic et al. used rotation forest [8] to improve classification accuracy for the gene expression data. Another approach – called *RFGS* (*random forest gene selection*) – was proposed by Chen et al. to identify biologically relevant genes concerning *leukemia* and *prostate* cancers from gene expression profiles [9]. Furthermore, aggregation of *SVM* classifiers, if coupled with the *random subspace* technique [10], also improves classification of microarray data. The second group of approaches to the process of gene selection and classification of microarray data is hybrid classifiers. For instance, in [11] the new *SVST* (*support vector sampling technique*) method was proposed. The technique developed by *Chen and Lin* was based on identifying principal genes and their further usage to cancer classification by machine learning methods like *SVM* (*support vector machine*) or *BPNN* (*back-propagation neural network*).

Application of the *SVST* approach allowed to achieve significant improvement of classification results, including average 2–3 % better performance for *leukemia* and 6–7 % better performance for *prostate* cancer. Also Huerta et al. showed the results from the research on using the *GA* (*genetic algorithm*) in combination with *LDA* (*linear discriminant analysis*). The algorithm was able to obtain high prediction accuracy with selecting a small number of selected genes simultaneously [12]. Ghorai et al. achieved similar promising results in cancer classification from gene expression data by using the combination of filter and wrapper methods. It was shown in [13] that such approach allows reducing dimensionality and achieving efficiency comparable with the typical wrapper algorithms.

### 7.1.2 MicroRNA

Recently, research on *microRNA* (further called *miRNA*) has received a lot of attention, mainly because of its role in gene regulation [1]. The *miRNAs* are small noncoding *RNAs* of 20–24 nucleotides, usually excided from 60 to 110 nucleotide foldback *RNA* precursor structures [14]. The primary and one of the most comprehensive repositories of *miRNA* sequences and annotations is available online at the *Sanger Institute* [15]. The current release (*miRBase 21* from *June 2014*) contains 28,545 entries representing hairpin precursor *miRNAs*, expressing 35,828 mature *miRNA* products, in 223 species. The biogenesis of the *miRNAs* involves a complex protein system, which includes members of the *Argonaute* family, *Pol II*-dependent transcription, and the *RNase III*s *Drosha* and *Dicer* [16]. The *miRNAs* are involved in crucial biological processes, containing development, differentiation, apoptosis, and proliferation [17] through imperfect pairing with target messenger *RNAs* (i.e., *mRNAs*) of protein-coding genes and the transcriptional (or posttranscriptional) regulation of their expression. Many human *miRNAs* appear to influence diseases. Various research studies in the field of computational prediction of *miRNA* targets proved that at least 30 % of the human genes could be targeted by *miRNAs* [14, 18]. For instance, many *miRNAs* exist in genomic regions associated with cancers [19]. Moreover, in [20] it has been suggested that:

- *MiRNA* profiles (for mature and precursor *miRNAs*) of tumors differ significantly from the normal cells profiles of the same tissue.
- Different types of cancers are associated with the different *miRNA* expression patterns.

Braun et al. compared the expression levels of more than 200 human *miRNAs* in tumor and adherent tissues of more than 60 patients with seven types of cancer (*lung, colon, breast, bladder, pancreatic, prostate,* and *thymus* cancers). Then, they proved that each of these various cancers has different *miRNA* profile. In conclusion, the *miRNAs* are potential biomarkers for diagnosis of tumors. Furthermore, they may provide novel approaches to develop better medicines to cure these deadly diseases. Nonetheless, the *miRNAs* which assert more knowledge (information) about the differences between the normal and tumor tissues have not been fully defined yet [21]. Thus, a lot of research on different machine learning methods of data mining and classification of *miRNA* data sets were conducted in the last decade. The new algorithms of *miRNAs* analysis and classification are based on using different machine learning approaches, including (*i*) *unsupervised* methods (e.g., *artificial neural networks*), (*ii*) *semi-supervised* methods (e.g., new methods discussed in [22]), and (*iii*) *supervised* methods (e.g., *decision rules, decision trees*). The list of a few efficient approaches to *miRNAs* by using various machine learning methods includes:

- Analysis based on using *ANN* (*artificial neural network*) to identify *miRNAs* associated with specific *breast cancer* phenotypes [23].
- Improving the quality of cancer classification by adaptation of *semi-supervised* machine learning algorithms [22].
- Comparison of *SVM* (*support vector machine*) classification and analysis of *miRNA* expression profiles with other machine learning approaches [21].
- Mining *miRNA* expression data (*decision rule* induction) using *rough set* methodology [24].

For instance, research on identification of *miRNA* signatures which allows to predict a few receptor status of *breast* cancer patients provided an insight into the regulation of *breast* cancer phenotypes and progression that were described by Lowery et al. in [23]. The artificial neural network analysis identified biologically relevant *miRNAs* related with specific *breast* cancer phenotypes. Thus, authors proved that relationship between selected *miRNAs* and *estrogen, progesterone,* and *HER2/neu* receptors status shows a role for these *miRNAs* as potential biomarkers in the process of *breast* cancer classification. In [22] two *semi-supervised* machine learning approaches – called *self-learning* and *co-training* – were adapted to enhance the quality of cancer subtype classification, based on both *miRNA* and gene expression profiles. These two newly developed approaches (*self-learning* and *co-training*) had better classification accuracy than *RF* (*random forest*) and *SVM* (*support vector machine*) baseline classifiers. Moreover, often used as one of the most effective algorithms for analysis of *miRNA* expression profiles are algorithms based on *supervised* machine learning approach. For instance, in [21] the computational method based on using *SVM* to analysis of *miRNA* expression profiles from wet lab experiment was presented. In addition, the effective, two-step feature selection procedure has been proposed. By using a combination of *Fisher* criterion and *SVM* with linear kernel, Tran et al. founded *miRNAs* which allowed to differentiate normal vs. tumor cells. Based on the tests performed for the *microarray* data sets (with expression of human *miRNAs*), it was proved that the developed *SVM*-based classifier outperformed other machine learning approaches, such as *BPNN* (*back-propagation neural network*). Another good result of mining *miRNA* expression data – by the *LERS* (*Learning from Examples based on Rough Sets*) classification system – was described in [24]. In the performed research, *Fang and Grzymała-Busse* used *miRNA* expression level data set describing 11 types of human cancers [1], whereas the main tool of classification was *MLEM2* (*Modified*

*Learning from Examples Module version 2*) algorithm of the *LERS* system. The induced set of decision rules classified correctly all but one cases of *breast* cancer and all cases of *ovary* cancer from the testing data set, which contained 17 poorly differentiated cases of four various types of cancers. Analysis of the induced set of decision rules showed that the prediction of these two types of cancers was based on using seven *miRNA* expression levels, three *miRNAs* with well-known functions and uncovered strong connections to the certain type of tumors and four *miRNAs* with some new functions (i.e., not determined experimentally before).

However it should be noticed that mining microarray data (including *miRNA* data sets) may be problematic. One of the most important problems is the small number of cases (e.g., less than 100) compared with the huge number of features (attributes) equal to several hundreds or even tens of thousands. Moreover, usually only a few selected genes from all saved in the microarray data set are essential for the examined disease. These facts may cause instability of newly created (induced) classifiers, as well as incorrect selection of features, described by so-called redundant attributes characterized by a low correlation with the decision. In addition, the huge number of features in the learning data set (*microarrays*) very often entails increasing complexity of the created classifiers, containing (*i*) many useless (redundant) conditions and (*ii*) many over-fitted, too specific, and complex decision rules.

### 7.1.3 Limitations of Classifiers

The development of classifiers used for mining microarray data imposes the necessity of considering a few issues, including:

- Ensuring the high, acceptable level of classifiers' stability.
- Selecting only features (attributes) strongly correlated with the decision.
- Avoiding too much complexity of the developed classifier.

At first, the instability of classifiers is usually caused by using the small number of cases available during the learning process. Secondly, the microarray data sets can contain samples described by the huge number of features (attributes) equal to several hundreds or even tens of thousands. In brief, if data are highly dimensional and the cardinality of training data set is small compared to data dimensionality, then the newly created (induced) classifier will be weak and inefficient. The unstable classifiers are characterized by the high value of variance. Thus, such classifiers have high classification accuracy of the cases from the learning data set only (equals even 100 %), but high error rate for the cases from the testing data set (or for new, unseen cases) at the same time. Therefore, to decrease the error rates and achieve acceptable values of the classification, the additional methods, for example, *bagging*, *boosting*, or *RSM* (*random subspace method*), may be applied. These approaches (so-called *ensemble* methods) are based on the process of aggregating classifiers, in combination with the voting (weighted or unweighted) procedure. The *bagging* (i.e., *bootstrap aggregating*) was proposed in [25] to improve the stability and accuracy of machine learning algorithms in statistical classification and regression. In addition bagging reduces the variance and prevents over-fitting of the newly created classifiers. Bagging is used mainly for improving unstable classifiers such as artificial neural networks or decision trees. For instance, *Breiman* proposed the *RF* (*random forest*) classifier consisted of a collection of decision tree structure-based classifiers. The second group of methods used for improving performance of any (even weak) machine learning algorithm is based on *boosting* approach. The first polynomial-time boosting algorithm was proposed in *1989* by Schapire [26]; then 1 year later *Freund* developed much more efficient boosting algorithm [27], but optimal only in a certain sense and characterized by some practical drawbacks. One of the most efficient boosting algorithms is *AdaBoost* (*adaptive boosting*), first time presented in *1995* [28]. *Freund and Schapire* succeeded in obtaining performance of the final (*aggregated*) classifier better than the performance of any of the *baseline* classifiers. Furthermore, *AdaBoost* algorithm solved many of practical difficulties of the earlier boosting algorithms [29]. The third approach of aggregating machine learning models is *RSM* (*random subspace method*) [30] that can be used for improving the performance of various types of classifiers, e.g., *decision trees*, *SVM* (*support vector machine*), or *kNN* (*k-nearest neighbor*). In [31], the author presented the aggregated classifier (*decision forest*) consisting of multiple decision trees, constructed in randomly chosen subspaces. Recently, the methods of combining classifiers are increasingly used in the research on gene expression data analysis. For instance, in [32] a new *ensemble* machine learning algorithm for classification and prediction of gene expression data was presented. The *aggregated* classifiers allowed using learning data sets containing much smaller number of cases than the algorithms creating only single (*baseline*) classifiers. Moreover, finding an ordinary (i.e., single) classifier that works so effectively as an aggregated (combined) classifier is very difficult or even impossible. Thus, in our research a new approach, based on inducing an aggregative classifiers – called decision *rule generations* – instead of inducing only single (baseline) set of decision rules was used (described in details in subchapter 7.2 "Learning from Examples Based on Rule Generations").

The second problem of mining microarray data (or other large data sets) is caused by the huge number of attributes (hundreds or thousands) describing each case saved in such data set, whereas only a few selected genes (attributes) are essential for any examined disease. Thus, the feature (attribute) selection must be conducted carefully. The process of building an efficient classifier has to be based on rejecting many redundant attributes characterized by low correlation with the dependent variable (decision), and selection of only a few dominant attributes strongly correlated with a decision at the same time. For instance, finding relations between attributes to eliminate redundant information can be accomplished by using some elements of the *rough set* theory [33]. In a case of decision rule-based classifiers, the high efficiency of feature selection and high value of further classification accuracy are achieved for so-called local decision rule induction algorithms. One of the most effective of local type algorithms is *LEM2* (*Learning from Examples Module version 2*) [34]. In the *LEM2* algorithm, the decision rules are induced by using the specific approach, based on exploring the search space of blocks of *A-V* (*attribute-value*) pairs, instead of performing calculations on partitions of the entire data set like in the global covering approaches. According to that fact, in our research the *LEM2* algorithm and its further modified version *MLEM2* (*Modified LEM2* [35]) were used as one of the component classifiers (i.e., base models) in the procedure of decision *rule generation* induction.

The last limitation of newly induced decision rule-based classifiers is related to the complexity of such learning model, despite of the simplicity of this formalism of knowledge representation. Difficulty of interpretation of such learning model can be caused by (*i*) too many decision rules and (*ii*) complexity of selected decision rules, containing too many attributes. Optimization of decision rules usually involves the detection of anomalies related to:

- Redundant knowledge (i.e., multiplied and absorbed decision rules).
- Conflicting decision rules (i.e., containing equivalent conditions, but different decisions).
- Missing decision rules (i.e., logical combinations of allowed input values, not covered yet by any decision rule).

Verification of correctness and consistency of previously induced decision rules is usually performed by using supplementary (external) computer systems. For instance, *COVERAGE* system allows to verify decision rules induced for multi-agent architectures. The second example – *VALENS* (*VALid ENgineering Support*) system – provides verification and validation of decision rules. Furthermore, some decision rule induction algorithms – as the *LEM2* algorithm, implemented in *LERS* classification system – have built-in procedures of verification of decision rules, including removing (*i*) redundant rules from the output set of decision rules and (*ii*) redundant conditions of the decision rules [36].

## 7.1.4 Imperfections of Data Sets

In the process of creating new classifiers beyond the aspects related to the specific construction of microarray data, also the imperfections of typical data sources must be considered. Typical real data sets could be (*i*) affected by errors, (*ii*) described by *numerical* attributes (so-called *continuous* attributes), (*iii*) *incomplete*, i.e., some attribute values of selected cases are lost, and/or (*iv*) *inconsistent*, i.e., at least two (or more) cases are characterized by the same attribute values but different decision values. Below methods of handling such imperfect data sets for induction of decision rule-based classifiers are discussed.

For instance, in Table 7.1 the first case has value 42 for the *Weakness* attribute, when this attribute is *symbolic* with only two permissible values *yes* and *no*. Such error(s) has to be corrected before creation of classifier, i.e., induction of decision rules.

7 LEMRG Algorithm

**Table 7.1** An example of the *inconsistent* and *incomplete* data set (*decision table*), with some *numerical* attributes

| Case | Attributes | | | | Decision |
|---|---|---|---|---|---|
| | Temperature | Headache | Weakness | Nausea | Flu |
| 1 | 41.6 | yes | 42 | no | yes |
| 2 | ? | yes | no | yes | yes |
| 3 | 37.0 | no | no | ? | no |
| 4 | 37.0 | ? | yes | yes | yes |
| 5 | 38.8 | no | yes | no | yes |
| 6 | 40.2 | no | no | no | no |
| 7 | 36.6 | no | yes | no | no |
| 8 | 36.6 | no | yes | no | yes |

Extended from [37]

The second problem is related to data described by *numerical* attributes, for instance, *Temperature* is represented by real numbers. Thus, before (or during) decision rule induction, the process of converting numerical attributes into symbolic ones – called *discretization* or *quantization* – must be performed. The process of conversion is fulfilled by partitioning of the several numerical attribute domains into selected intervals. The most important point is the maximization of the classification accuracy of the decision rule set induced from such quantized data set. The most commonly used techniques of discretization are *equal interval frequency*, *equal interval width*, *minimal class entropy*, *minimum description length*, or *clustering* [38].

Another problem during the process of learning classifier can be caused by *incomplete* data (with *lost* values). In Table 7.1, lost values are denoted by *?*. For such data sets are two different approaches to handling missing attributes values, called *sequential* and *parallel*. In the sequential approach, called also preprocessing, the following methods are used: (*i*) ignoring (deleting) cases with missing attribute values or (*ii*) replacing missing attribute values by the most common value. The other sequential methods are (*i*) assigning all possible values to the missing attribute values, (*ii*) replacing a missing attribute value by the mean (for numerical attributes), (*iii*) assigning to a missing attribute value the corresponding value taken from the closest fit case, and/or (*iv*) replacing a missing attribute value by a new value, computed from a new data set, considering the original attribute as the decision [39]. It should be noticed that there is no best, universal method of handling missing attribute values. In other words, for the specific data set, the best method of handling missing attribute values should be chosen individually, using as the criterion of optimality the arithmetic mean of many multifold cross validation experiments.

The last possible problem with input data sets relates to *inconsistency*. In Table 7.1, cases 7 and 8 are conflicting with each other. Often such situation occurs in the data sets with missing attribute values. One of the most known approaches of handling inconsistent data is based on using *rough set* theory [40]. Conflicting cases are not removed from the data set, but all concepts are approximated by calculating new sets, called *lower* and *upper* approximations.

## 7.1.5  Supervised Machine Learning (Decision Rules)

Machine learning models can be saved in different forms, including:

- *Symbolic* (e.g., *decision rules, decision trees*) – called *knowledge-oriented* methods.
- *Nonsymbolic* (e.g., *artificial neural networks*), known as "*black box*" methods.

From a wide spectrum of knowledge representation, the most popular seems to be decision rules, usually expressed in the form *IF... THEN....* The knowledge saved in the form of *decision rules* (or *decision trees*) is easier for humans to comprehend, rather than for the other machine learning models. The decision rule-based knowledge representation is the result of the decision rule induction process, which is one of the fundamental tasks of machine learning. A selected decision rule allows to find a decision, based on the certain conditions. Representative list of classification systems [36], using knowledge saved in the form of decision rules (or other equivalent form), includes:

- Systems based on finding local or global coverings, like *AQ, CN2, GTS, LEM2*, and *PRISM*.
- Systems using rough set theory (e.g., *LERS* or *Rosetta*).
- Systems inducing decision rules by using *belief networks* (e.g., *BeliefSEEKER* system), *ANN* (*artificial neural networks*), *genetic algorithms*, or *clusters analysis*.

According to [41], decision rules can be categorized into three groups, including (*i*) *classification* rules, (*ii*) *characteristic* rules, and (*iii*) *association* rules. Induction of decision rules belongs to the *supervised* machine learning, i.e., all cases are pre-classified by an expert, by assigning the decision value for each case. The decision rules are induced from the data organized usually in the form of *decision table* such as Table 7.1, in which cases are described by *attributes* (independent variables) and the *decision* (dependent variable). In the example of decision table presented in Table 7.1, attributes are *Temperature, Headache, Weakness*, and *Nausea*, whereas decision is *Flu*, and the set of all cases with the same decision value is called a *concept*. Thus, the decision table from an example contains two concepts (classes), including:

- Case set {*1, 2, 4, 5, 8*} is a concept of all cases affected by flu (i.e., for each case from this subset, the corresponding value of decision *Flu* is *yes*).
- Case set {*3, 6, 7*} is a concept of all cases not affected by flu (i.e., for each case from this subset, the corresponding value of decision *Flu* is *no*).

In principle, decision rules are induced directly from the learning data set or indirectly, due to the converting of other formalisms of knowledge. For instance, one of the most common indirect algorithms is *C4.5* [42], which is an extended version of *ID3* algorithm [43], based on mapping all various paths of the decision tree (i.e., from root to all leafs) to the format of the decision rules (*IF... THEN...* notation). Among different algorithms of direct decision rule induction, the most popular are methods based on sequential covering of consecutive cases from the learning data set [44], whereas the simultaneous covering methods are used less frequently.

There exist many decision rule induction algorithms; therefore only a few of them will be mentioned. The decision rule induction algorithms can be classified generally as *global* or *local*. The example of representative algorithm of global decision rule induction algorithms is *LEM1* (*Learning from Examples Module version 1*). However, it should be noticed that in most cases, the local decision rule induction algorithms give better results of classification, rather than global ones [37]. Among many algorithms of local decision rule induction, one of the most often used is *LEM2* (*Learning from Examples Module version 2*) [34]. In brief, in the *LEM2* algorithm, a decision rule set is induced by using the specific approach, based on exploring the search space of blocks of *A-V* (*attribute-value*) pairs, instead of performing calculations on partitions of the entire data set like for the global covering approaches. The *LEM2* algorithm computes a local covering and then converts it into a set of discriminant decision rules, i.e., the smallest set of minimal decision rules describing each concept of the input data set. An example of decision rule induction process using *LEM2* algorithm is discussed in details in the further part of this chapter (see in Sect. 7.2.1 "*LEM2*"). The further well-known, efficient modifications of *LEM2* algorithm are *MODLEM* [45], *ELEM2* [46], and *MLEM2* [35]. For instance, the *MLEM2* (i.e., *Modified LEM2*) algorithm is a newer, modified version of the LEM2 algorithm, which allowed to induce decision rule sets for data sets which are incomplete, inconsistent, and described by continuous (numerical) attributes [47]. In the case of incomplete data for any attribute with missing values, the *A-V* (*attribute-value*) blocks are computed only for the existing values of such attributes. Moreover, through the use of *rough set* theory – i.e., by computation of *lower* and *upper* approximation for each concept from the whole set – decision rules can be induced for the inconsistent data sets, i.e., containing cases which may conflict each other. Both lower and upper approximations can be calculated in a different way, while usually one of the three well-known methods – called *singleton*, *subset*, and/or *concept* type approximation [39] – is used. It should be noticed that in a case of inconsistent data set, two separated sets of decision rules are induced, including:

- *Certain* decision rules – induced for the *lower* approximation (of each concept).
- *Possible* decision rules – induced in the same manner, but for the *upper* approximation (of each concept) [39].

**Table 7.2** Error rates of different decision rules and decision trees for selected data sets [50]

| Data set | LERS | AQ15 | C4.5 |
|---|---|---|---|
| Lymphography | 19 % | 18–20 % | 23 % |
| Breast cancer | 30 % | 32–34 % | 33 % |
| Primary tumor | 67 % | 59–71 % | 60 % |

For numerical attributes, *MLEM2* algorithm first sorts numerical values in the ascending order. Then, the cut points as averages for any two consecutive values of the sorted list of attribute values are computed. For instance, the *cut points* of *Temperature* from Table 7.1 are 36.8, 37.9, 39.5, and 40.9. Afterward, for each cut point, two blocks are created as follows:

- The first block contains all cases for which values of the selected (numerical) attribute are smaller than selected cut-point value.
- The second block contains remaining cases (i.e., for which values of the selected, numerical attribute are greater than selected cut point value).

For instance, the all block values for *Temperature* from Table 7.1 are as follows (36.6...36.8), (36.8...41.6), (36.6...37.9), (37.9...41.6), (36.6...39.5), (39.5...41.6), (36.6...40.9), and (40.9...41.6). Further, the search space of *MLEM2* algorithm is the set of all blocks calculated this way and block of remaining (*symbolic*) attributes (in an example from Table 7.1, these attributes are *Headache*, *Weakness*, and *Nausea*). Finally, from that point, decision rules are induced in *MLEM2* in the same way like for *LEM2*.

Both – *LEM2* and *MLEM2* – algorithms are characterized by the high efficiency in classification process of various types of data, mainly with reference to classification accuracy. Comparison of the *LERS* (*Learning from Examples based on Rough Sets*) classification system, which includes implementation of *LEM2* and *MLEM2* algorithms, with other well-known classification systems (algorithms) can be found in Table 7.2. Namely, for all of three tested medical data sets, *lymphography*, *breast cancer*, and *primary tumor* from the *UCI Repository* (*University of California, Irvine*) [48] performance of the *LERS* system were fully comparable with performance of other well-known decision rules (or decision trees) induction systems, i.e., *AQ15* [49] and *C4.5* [42].

Another efficient decision rule induction algorithm is *GTS* (*General-To-Specific*) [51]. However that algorithm is able to induce decision rules only for complete and consistent data sets, and further all attributes must be symbolic.

One of the most important measures of quality of decision rule induction algorithms is the *error rate*. Estimation of misclassification is dependent on the number of cases contained in the training data set. One of three well-known methods, including *leave-one-out*, *n-fold cross* validation, and *hold-out*, is used [52], depending on the data set cardinality as follows:

- *Leave-one-out* method for data sets with less than 100 cases.
- *n-fold cross* validation for data sets which contain from 100 to 1,000 cases.
- *Hold-out* validation for the larger data sets, with more than 1000 cases.

## 7.2 Learning from Examples Module Based on Rule Generations

The following section describes a newly developed *LEMRG* algorithm. Our algorithm is designed for mining *miRNA* data sets, mainly to improving *stability* and, consequently, *accuracy* of classification for the newly created machine learning models. The instability of newly created classifiers is caused by the small number of cases in the learning data sets, which is typical for *multiarray*s (e.g., for *miRNA* data sets). To overcome this limitation, a few steps are undertaken: (*i*) creation of many *baseline* classifiers and then (*ii*) *aggregation* of them, rather than creating only a single classifier. The main concept of our novel approach to data mining is based on iteratively inducing many subsequent decision rule sets – called decision *rule generations* (Fig. 7.1) – instead of inducing a single decision rule set, as it is done routinely.

As the baseline models (classifiers), many different machine learning models, including *symbolic* (e.g., *decision trees*) and *sub-symbolic* (e.g., *artificial neural networks*), can be applied. It should be noticed that the overall approach of creating, and further using aggregated classifiers (e.g., *random forest* vs. *single decision tree*), is not new. However, the novelty of our approach is related to using decision rules as the baseline models (classifiers), in addition with reference to classification of

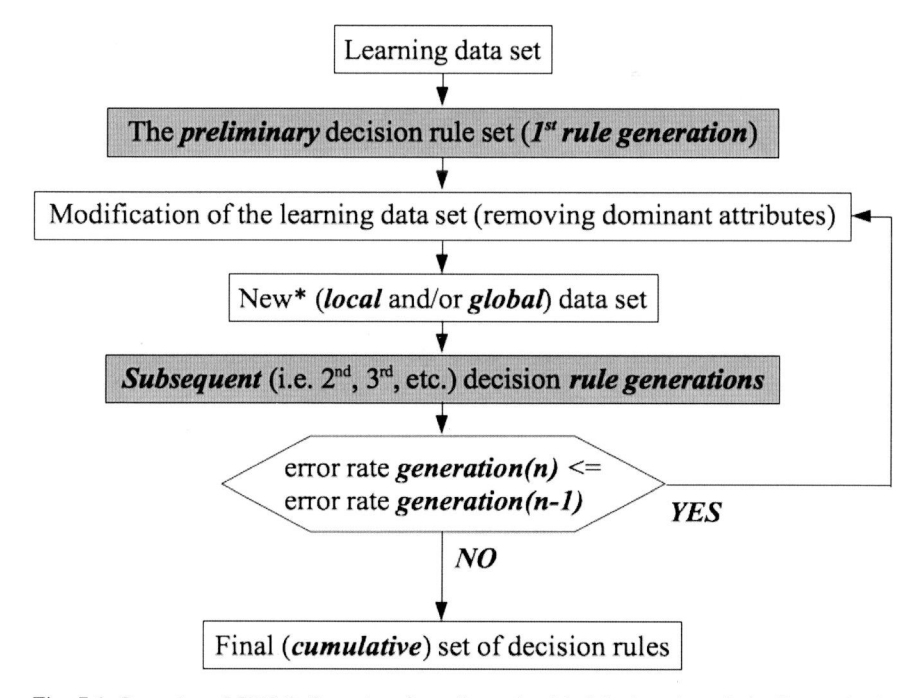

**Fig. 7.1** Procedure *LEMRG* (*Learning from Examples Module based on Rule Generations*). Induction of the decision *rule generations* (with the gray color marked modules of inducing subsequent decision rule sets, i.e., subsequent baseline classifiers)

*miRNA*s data sets, not previously described (and/or explained) in details in the literature. The choice of the decision rules as a baseline model was caused by a few reasons. At first, the decision rule-based knowledge representation is easier for humans to comprehend, than other learning models, when classification accuracy is similar (or even better) to the other machine learning approaches. That feature can be clearly noticed primarily in comparison with *sub-symbolic* knowledge representation, called *black box* methods (e.g., *artificial neural networks* and/or *statistical classifiers*). Furthermore, the induction of decision rules is also a fundamental tool of data mining among symbolic machine learning models (e.g., *decision trees*). Regularities hidden in the analyzed data, expressed in terms of decision rules, can be directly implemented in the specialized classification expert system. On the other hand, the generation of decision trees requires performing the translation of such learning models at first by converting all paths from root (main node) to leafs (decision nodes).

In the novel approach (Fig. 7.1), the first decision *rule generation* is induced in a normal manner, from the entire learning data set. Next, in the process of inducing subsequent sets of decision rules (i.e., decision *rule generations*), the specific approach is used. After each iteration of inducing single set of the decision rules (i.e., subsequent decision *rule generations*), the learning data set is modified, by applying concepts of so-called (*i*) *local* and (*ii*) *global* data sets. The *LEMRG* algorithm of rule generation induction is divided into several subtasks (Fig. 7.1) as follows:

1. Induction of the *first* (preliminary) decision *rule generation*.
2. Modification of the learning data set.
3. Induction of *subsequent* decision *rule generations*, in combination with further modification of learning data set(s), as long as the classification error rate of the next decision rule generation is smaller (or equal at least) than for the previous ones.
4. Process of merging decision rules from the subsequent decision *rule generations* into the final (*cumulative*) set of decision rules.

In the performed research, induction of the subsequent generations of decision rules was accomplished by using *LEM2* and *MLEM2* algorithms, mainly because of the high value of classification accuracy of various types of data. Rules induced by using *LEM2* (and *MLEM2*) algorithms contain attribute-value pairs called as *dominant pairs*, which unlike the *redundant* attributes have a high value of correlation with all values of the decision.

In the next sections, selected modules of *LEMRG* algorithm will be described in details, showing:

- Process of inducing baseline classifier in the form of single decision rule set, by an example of using *LEM2* algorithm.
- Modification of the source (initial) learning data set, by its division into two new learning subsets (*local* and *global*).

## 7 LEMRG Algorithm

**Table 7.3** An example data set (*decision table*) [37]

| Case | Attributes | | | | Decision |
|------|------------|---------|---------|--------|----------|
|      | Temperature | Headache | Weakness | Nausea | Flu |
| 1 | very_high | yes | yes | no | yes |
| 2 | high | yes | no | yes | yes |
| 3 | normal | no | no | no | no |
| 4 | normal | yes | yes | yes | yes |
| 5 | high | no | yes | no | yes |
| 6 | high | no | no | no | no |
| 7 | normal | no | yes | no | no |

### 7.2.1 LEM2

The *LEM2* algorithm allows induction of classification rules. Data sets from which such decision rules are induced are presented in the form of two-dimensional table (*decision table*), where cases are labeled for rows and variables are labeled as the *attributes* (independent values) and a *decision* (dependent value). Furthermore, *LEM2* belongs to *supervised* machine learning methods; thus all cases from the data set must be pre-classified by an expert (Table 7.3).

In the *LEM2* algorithm, a decision rule set is induced by using the specific approach, based on exploring the search space of blocks of *A-V* pairs. Its input data set is a *lower* (or *upper*) approximation of a concept [50], so its input data are always consistent. In brief, the *LEM2* algorithm computes a local covering and then converts it into a set of decision rules, i.e., the smallest set of minimal decision rules describing each concept of the input data set. To describe *LEM2* algorithm in details, some definitions from [34, 37] will be quoted below, as follows:

- $[t]$ – for the attribute-value pair $(a, v) = t$, a block of $t$ – denoted by $[t]$ – is a set of all cases from the entire data set $U$ such that for an attribute $a$ have value equal $v$.
- Set $B$ – is the nonempty *lower* (or *upper*) approximation of a concept represented by the decision-value pair $(d, w)$.

Thus, the set $B$ depends on the set $T$ (i.e., set of attribute-value pairs $(a, v) = t$) if and only if the conditions described by Eq. 7.1 are satisfied:

$$\emptyset \neq [T] = \bigcap_{t \in T} [t] \subseteq B \qquad (7.1)$$

Then, set $T$ is a *minimal complex* of $B$ if and only if $B$ depends on $T$ and there is not any adequate subset $T'$ (of set $T$) such that $B$ depends on subset $T'$. Let $\tau$ be the nonempty collection of nonempty sets of attribute-value pairs. Then, $\tau$ is a *local covering* of set B if and only if three following conditions are satisfied:

- Each member $T$ of $\tau$ is a minimal complex of $B$.
- $\cup_{T \in \tau}[T] = B$.
- $\tau$ is minimal, i.e., $\tau$ has the smallest possible number of members (decision rules in the final set).

The scheme of the *LEM2* algorithm for finding a single local covering is presented below.

```
Procedure LEM2 [37]
(
input: a set B,
output: a single local covering τ of set B
);
begin
G := B;
τ := ∅;
while G ≠ ∅
begin
T := ∅;
T(G) := {t | [t] ∩ G ≠ ∅};
 while T = ∅ or [T] ⊄ B
begin
 select a pair t ∈ T(G) such that | [t] ∩ G | is
             maximum;
             if a tie occurs, select a pair t ∈ T(G) with the
             smallest
             cardinality of [t];
             if another tie occurs, select the first pair;
T := T ∪ {t};
G := [t] ∩ G;
T(G) := {t | [t] ∩ G ≠ ∅};
T(G) := T(G) - T;
end {while}
for eacht ∈ Tdo
  if [ T - {t}] ⊆ BthenT := T - {t};
τ := τ ∪ {T};
G := B - ∪_{T∈τ}[T];
end {while};
for each T ∈ τdo
if ∪_{S∈τ-{T}}[S] = Bthenτ := τ -{T};
end {procedure}.
```

## 7 LEMRG Algorithm

For a set $X$, $|X|$ denotes the *cardinality* of $X$. The first step of the *LEM2* algorithm is to compute all *A-V* (*attribute-value*) pair blocks, where, for example, decision table (Table 7.3), these blocks are as follows:

[(Temperature, very_high)] = {1}
[(Temperature, high)] = {2, 5, 6}
[(Temperature, normal)] = {3, 4, 7}
[(Headache, yes)] = {1, 2, 4}
[(Headache, no)] = {3, 5, 6, 7}
[(Weakness, yes)] = {1, 4, 5, 7}
[(Weakness, no)] = {2, 3, 6}
[(Nausea, no)] = {1, 3, 5, 6, 7}
[(Nausea, yes)] = {2, 4}

Further the process of inducing decision rules for the concept $\{1, 2, 4, 5\}$ (i.e., *Flu = yes*) is presented. At first $B = G = \{1, 2, 4, 5\}$; thus the set $T(G)$ of all relevant attribute-value pairs is as follows:

{
(Temperature, very_high), (Temperature, high), (Temperature, normal),
(Headache, yes), (Headache, no),
(Weakness, yes), (Weakness, no),
(Nausea, no), (Nausea, yes)
}

The next step relates to identify attribute-value pairs $(a,v)$ which have the largest value of the $|[(a,v)] \cap G|$. The highest cardinality of the set $|[(a,v)] \cap G|$ equals to 3 have two selected *A-V* pairs from $T(G)$, in it *(Headache, yes)* and *(Weakness, yes)*. The next estimated criterion is the size of the attribute-value pair block. In our example $|[(Headache, yes)]| = 3$ is smaller than $|[(Weakness, yes)]| = 4$, so *(Headache, yes)* pair is selected. Furthermore, $[(Headache, yes)] \subseteq B$; thus *(Headache, yes)* is the first *minimal complex* of $G$. Then, the new subset G is equal to $B - [(Headache, yes)] = \{1, 2, 4, 5\} - \{1, 2, 4\} = \{5\}$. Further, the set $T(G)$ contains the following attribute-value pairs:

{
(Temperature, high), (Headache, no), (Weakness, yes), (Nausea, no)
}

At this point the first criterion – i.e., the highest value of the $|[(a,v)] \cap G|$ – is equal for all four attribute-value pairs. Then, the second criterion selected (*Temperature*, *high*) pair, because the |[(*Temperature, high*)]| = 3 is smaller than cardinality of the remaining *A-V* blocks. Nonetheless, [(*Temperature, high*)] = {2, 5, 6} $\not\subseteq B$, so the additional attribute-value pairs have to be added to the decision rule (i.e., additional iteration of the internal loop must be performed). Thus, the next candidates are (*Headache, no*) and (*Weakness, yes*), because of the fact that the size of their blocks is equal to 4 for both of these attribute-value pairs. Then, based on heuristics the (*Headache, no*) is selected. However [(*Temperature, high*)] ∩ [(*Headache, no*)] = {5, 6} $\not\subseteq B$ = {1, 2, 4, 5}, so the (*Weakness, yes*) attribute-value pair has to be added as well. Then, [(*Temperature, high*)] ∩ [(*Headache, no*)] ∩ [(*Weakness, yes*)] = {5} $\subseteq B$ = {1, 2, 4, 5}, so the set {(*Temperature, high*), (*Headache, no*), (*Weakness, yes*)} is the candidate for the minimal complex. Then, the built-in procedure of possible removing redundant conditions should be performed:

**for each** $t \in T$ **do**
 **if** $|T - \{t\}|$ **then** $T := T - \{t\}$

As the result, the second minimal complex is identified as {(*Temperature, high*), (*Weakness, yes*)}. Finally, the local covering of $B$ = {1, 2, 4, 5} is the set:

{{(*Headache, yes*)}, {(*Temperature, high*), (*Weakness, yes*)}}

The complete decision rule set, i.e., induced for all concepts of our example *decision table* (Table 7.3) by using *LEM2* algorithm, is as follows:

Rule no. 1: 1, 3, 3
(*Headache, yes*) → (*Flu, yes*) covered {1, 2, 4}

Rule no. 2: 2, 1, 1
(*Temperature, high*) & (*Weakness, yes*) → (*Flu, yes*) covered {5}

Rule no. 3: 2, 2, 2
(*Temperature, normal*) & (*Headache, no*) → (*Flu, no*) covered {3, 7}

Rule no. 4: 2, 2, 2
(*Headache, no*) & (*Weakness, no*) → (*Flu, no*) covered {3, 6}

In brief, the three numbers associated with each decision rule means, respectively, the total number(s) of (*1*) attribute-value pairs in the left-hand side of the decision rule (i.e., in the conditional part), (*2*) cases classified correctly by the decision rule during the learning process, and (*3*) training cases matching the left-hand side of the decision rule.

## 7.2.2 Local and Global Data Sets

One of the most important steps of the *LEMRG* algorithm is a modification of the training data set, performed for each of the subsequent decision *rule generations*. This modification is based on creation of two new data sets – so-called local and global – by using (*i*) previously existing learning data set – initial or modified in the previous step – and (*ii*) the set of decision rules induced in the current *rule generation* step. The process of modification of the training data set is based on the following steps:

- Creation of the *local* data set – only values from the learning data set, which satisfy decision rules, should be replaced by the *?*.
- Creation of the *global* data set – all values from the learning data set that correspond to any condition of any rule should be replaced by *?*, without checking a decision value, as well as any conditions of decision rules.

The above procedure allows to reject (i.e., *switch off*) from the learning data set only selected *A-V* pairs, instead of switching off all attributes (i.e., for all possible attribute values). Removing selected attribute-value pairs is intended to avoid induction of identical (or very similar) decision rules in the further decision *rule generations*.

An example of such modification process of the initial data set presented in Table 7.3 is shown in Tables 7.4 and 7.5. Let's check in detail the differences of converting original data set into the new (*i*) *local* and (*ii*) *global* data sets, for a single decision rule:

(*Headache, no*) & (*Weakness, no*) → (*Flu, no*).

**Table 7.4** The new *local* data set after application of one selected decision rule (*Headache, no*) & (*Weakness, no*) → (*Flu, no*) (the changed values, replaced by the *?* marked by the light-gray color)

| Case | Attributes | | | | Decision |
|---|---|---|---|---|---|
| | Temperature | Headache | Weakness | Nausea | Flu |
| 1 | very_high | yes | yes | no | yes |
| 2 | high | yes | no | yes | yes |
| 3 | normal | ? | ? | no | no |
| 4 | normal | yes | yes | yes | yes |
| 5 | high | no | yes | no | yes |
| 6 | high | ? | ? | no | no |
| 7 | normal | no | yes | no | no |

**Table 7.5** The new *global* data set after application of one selected decision rule (*Headache, no*) & (*Weakness, no*) → (*Flu, no*) (the changed values, replaced by the ? marked by the light-gray 1; color)

| Case | Attributes | | | | Decision |
|---|---|---|---|---|---|
| | Temperature | Headache | Weakness | Nausea | Flu |
| 1 | very_high | yes | yes | no | yes |
| 2 | high | yes | ? | yes | yes |
| 3 | normal | ? | ? | no | no |
| 4 | normal | yes | yes | yes | yes |
| 5 | high | ? | yes | no | yes |
| 6 | high | ? | ? | no | no |
| 7 | normal | ? | yes | no | no |

At first, for the *local* data set, only attribute values from the initial data set, which satisfy that decision rule, could be changed. Such attribute values will be *switched off* by putting the *?* mark instead of its original values for the selected cases. In our example some values of *Headache* and *Weakness* attributes should be changed only for cases satisfying jointly all of the following conditions:

- *Headache* attribute has value *no*.
- *Weakness* attribute has value *no*.
- The decision *Flu* is equal to *no*.

As the result, four values have to be changed, including values *no* of *Headache* and *Weakness*, only for cases 3 and 6; see Table 7.4.

On the other hand, for the *global* data set, attribute values from the source data set, which satisfy at least one condition from the left-hand side of any decision rule, are changed. According to this procedure the selected attribute-values of the following cases should be changed as follows:

- Value *no* of *Headache* attribute is replaced by *?* for cases 5 and 7, despite the fact that *Weakness* attribute values of these both cases are equal to *yes*.
- Value *no* of *Headache* attribute is replaced by *?* for case 5 although the decision *Flu* is equal to *yes*.

Thus, in our example all occurrences of *no* value for both – *Headache* and *Weakness* – attributes are replaced by the *?*, i.e., *switched off* in the process of inducing further decision *rule generations* (Table 7.5).

The final *local* and *global* data sets, i.e., after applying all four decision rules, are presented in Tables 7.6 and 7.7, respectively.

It is clear that even for such trivial decision table, the contents of both data sets are different. For instance, in a case of *global* data set (Table 7.7) in the further

# 7 LEMRG Algorithm

**Table 7.6** The new *local* data set after application of the entire set of decision rule, i.e., four decision rules (the changed values, replaced by the *?* marked by the light-gray color)

| Case | Attributes | | | | Decision |
|---|---|---|---|---|---|
| | Temperature | Headache | Weakness | Nausea | Flu |
| 1 | very_high | ? | yes | no | yes |
| 2 | high | ? | no | yes | yes |
| 3 | ? | ? | ? | no | no |
| 4 | normal | ? | yes | yes | yes |
| 5 | ? | no | ? | no | yes |
| 6 | high | ? | ? | no | no |
| 7 | ? | ? | yes | no | no |

**Table 7.7** The new *global* data set after application of the entire set of decision rule, i.e., four decision rules (the changed values, replaced by the *?* marked by the light-gray color)

| Case | Attributes | | | | Decision |
|---|---|---|---|---|---|
| | Temperature | Headache | Weakness | Nausea | Flu |
| 1 | very_high | ? | ? | no | yes |
| 2 | ? | ? | ? | yes | yes |
| 3 | ? | ? | ? | no | no |
| 4 | ? | ? | ? | yes | yes |
| 5 | ? | ? | ? | no | yes |
| 6 | ? | ? | ? | no | no |
| 7 | ? | ? | ? | no | no |

decision rule induction process, only two of four original attributes will be used, in it:

- *Temperature* – with only one possible value equals *very_high*.
- *Nausea* – with two possible values *yes* and *no*.

At the same time, the two remaining attributes (*Headache* and *Weakness*) are completely *switched off*,. Whereas for the local data set (Table 7.6), all four attributes (for the remaining attribute-value pairs) still could be used in the induction of the further decision *rule generations*.

## 7.3 Results of Mining MiRNA Expression Data Using Rule Generations

The following section describes the results of mining the *miRNA* data set describing danger of human cancer appearance [1], by using our newly developed *LEMRG* (*Learning from Examples Module based on Rule Generations*) algorithm [2].

### 7.3.1 Data Sets

The data set used in our research was reported by Lu et al. in a study using a *miRNA* expression level for classifying human cancers [1]. The authors profiled 217 mammalian *miRNA* from 334 samples, including multiple human cancers, by using *bead-based flow cytometry* technique [1]. By applying a *Gaussian* weight-based nearest neighbor 11-class classifier, trained on a learning set of 68 more different tumors, Lu et al. were able to classify 17 poorly differentiated cases from testing data set, with an accuracy of 11 out of 17 correct. Thus, in our research on induction and validation of decision *rule generations*, two such *miRNA* data sets, i.e., learning and testing, were used. The *training* data set had 68 cases, each described by 217 attributes. Cases were distributed among 11 concepts, including *bladder* cancer (*BLDR*), *breast* cancer (*BRST*), *colon* cancer (*COLON*), *kidney* cancer (*KID*), *lung* cancer (*LUNG*), *melanoma* (*MELA*), *mesothelioma* cancer (*MESO*), *ovary* cancer (*OVARY*), *pancreas* cancer (*PAN*), *prostate* cancer (*PROST*), and *uterus* cancer (*UT*). The *testing* data set contained only 17 poorly differentiated cases, with the same set of 217 attributes (like for a learning data set) and only four concepts, including *colon* cancer (*COLON*), *ovary* cancer (*OVARY*), *lung* cancer (*LUNG*), and *breast* cancer (*BRST*).The number of cases divided among selected concepts (i.e., different cancer types) for both data sets is placed in Table 7.8.

Furthermore, we restricted our attention only to two concepts (*COLON* and *OVARY*), because for these two concepts the newly proposed approach provided the best results (Table 7.9).

### 7.3.2 Induction of Rule Generations

As it was mentioned above, in the newly approach to data mining, instead of inducing a single decision rule set – as it is done routinely – many decision rule sets (called decision rule generations) were induced. The *first* decision *rule generation* was induced in an ordinary way, from the whole learning data set consisted of all 68 cases (with all original values) with using *MLEM2* algorithm. Thus, 11 decision rules, one decision rule per each concept, were induced, where decision rules describing two classes of interest (*COLON* and *OVARY*) were as follows:

**Table 7.8** The number of cases of selected concepts for *learning* and *testing* data set(s)

| Concept (cancer type) | Number of cases | |
|---|---|---|
| | Learning data set | Testing data set |
| BLDR | 6 | – |
| BRST | 6 | 5 |
| COLON | 7 | 1 |
| KID | 4 | – |
| LUNG | 5 | 8 |
| MELA | 3 | – |
| MESO | 8 | – |
| OVARY | 5 | 3 |
| PAN | 8 | – |
| PROST | 6 | – |
| UT | 10 | – |

**Table 7.9** The number of correctly classified cases of *COLON* and *OVARY* concepts by using our new decision *rule generation* approach

| Decision rule set | Number of correctly classified cases | |
|---|---|---|
| | COLON | OVARY |
| *First* decision rule generation | 0 | 2 |
| *Second* decision rule generation | 1 | 1 |
| *Cumulative* set of decision rules (i.e., decision rules combined from the *first* and *second* generations) | 1 | 3 |

```
2, 7, 7
(EAM238, 6,724465…11,9952) & (EAM298, 9,33032…11,5169) →
(Label, COLON)
3,5,5
(EAM233, 5…6,87862) & (EAM208, 9,10495…11,7605) & (EAM342,
8,41185…10,8687) → (Label, OVARY)
```

Furthermore, the original learning data set was modified into two new data subsets, i.e., *local* and *global*. Then, the subsequent sets of decision rules – i.e., the *second* decision *rule generation* – were induced, separately for *local* and *global* data set(s). The *second decision rule generation* (only for the *global* data set) restricted to only *COLON* and *OVARY* concepts was as follows:

```
3, 7, 7,
(EAM358, 5...5,16315) & (EAM241, 5,78525...9,5293) & (EAM232,
9,14248...11,4941) → (Label, COLON)
3, 5, 5,
(EAM317, 5...5,1027) & (EAM225, 5,125...9,1908) & (EAM203,
7,1618...9,421175) → (Label, OVARY)
```

The process of induction of subsequent rule generations continues until considerable reduction of the quality of a new decision rule generation. In this example the classification accuracy of the *third* decision *rule generation* was much worse than for the *second* decision *rule generation*, so no further subsequent decision rule generations were induced.

### 7.3.3 Cumulative Rule Sets

Finally, the decision rules from the subsequent decision *rule generations* (i.e., *first* and *second* decision *rule generations*) were gradually collected into the new so-called *cumulative* decision rule set. This set of decision rules, limited to *COLON* and *OVARY* concepts, was as follows:

```
2, 2, 2,
(EAM238, 6,724465...11,9952) & (EAM298, 9,33032...11,5169) →
(Label, COLON)
3, 1, 1,
(EAM358, 5...5,16315) & (EAM241, 5,78525...9,5293) & (EAM232,
9,14248...11,4941) → (Label, COLON)
3, 2, 2,
(EAM233, 5...6,87862) & (EAM208, 9,10495...11,7605) & (EAM342,
8,41185...10,8687) → (Label, OVARY)
3, 1, 1,
(EAM317, 5...5,1027) & (EAM225, 5,125...9,1908) & (EAM203,
7,1618...9,421175) → (Label, OVARY)
```

It should be noticed that the decision rule *strengths* (marked by the second number associated with each decision rule) were changed. All decision rules from the first decision rule generation have rule strengths twice larger than decision rules from the second decision rule generation. Such procedure determines the order of using individual decision rules in the further process of classification. During the classification of new, unseen cases, the decision rules with the higher strengths are used earlier than the others, i.e., decision rules from the *first* decision *rule generation* are used before decision rules from the *second* decision *rule generation*, decision rules from the *second* decision *rule generation* before decision rules from the *third* decision *rule generation*, and so on.

The new classifier, saved in the form of a *cumulative* decision rule set, had better classification accuracy than the *single* (preliminary) decision rule set induced in the ordinary form (Table 7.9). For the *first* decision *rule generation* induced in the traditional way, the case of *colon* cancer was misclassified, and only two (out of three) cases of *ovary* cancer were classified correctly. While, for the final (*cumulative*) set of decision rules, all cases from both concepts – i.e., one case of *COLON* and three cases of *OVARY* – were classified correctly.

Furthermore, we compared the classification accuracy of *LEMRG* with other machine learning models. Implementations of all used algorithms were taken from the *WEKA* (*Waikato Environment for Knowledge Analysis*) machine learning toolkit developed at the *Waikato University, New Zealand* [53]. In our research 16 different classifiers were tested:

- *Naive Bayes* learner, strong and naturally resistant to noise in data.
- *KStar*, an instance-based classifier using entropy-based distance function.
- Decision rule-based classifiers: (*a*) *DecisionTable*, an implementation of *DTMaj* (*Decision Table Majority*) algorithm; (*b*) *JRip*, a propositional rule learner, *Repeated Incremental Pruning to Produce Error Reduction* (*RIPPER*), proposed as an optimized version of *IREP* algorithm; (*c*) *OneR*, a simple classifier which extracts a set of decision rules by using only a single attribute; (*d*) *PART*, using indirect approach for generating decision list, by using the C4.5 algorithm; (*e*) *ZeroR*, an 0-R classifier predicting the mean for a numeric class or the mode for a nominal class.
- Decision tree-based classifiers, including aggregated: (*a*) *DecisionStump*, for building and using decision stump, usually used in conjunction with a boosting algorithm; (*b*) *J48*, an implementation of the *C4.5* algorithm in *WEKA* toolkit; (*c*) *RandomTree*, for constructing a tree that considers *K* randomly chosen attributes at each node; (*d*) *REPTree*, a fast decision tree learner; (*e*) *random forest*.
- Metatype classifiers: (*a*) *AdaBoostM1* method; (*b*) *bagging*, for bagging the classifier to reduce the variance; (*c*) *classification* via *regression*, a multi-response linear regression, where class is binarized and one regression model is built for each class value; (*d*) *RSM*, a random subspace method, with *REPTree* baseline classifier.

It was found that only 3 out of all 16 tested classifiers enabled so effective classification as our newly developed approach (Table 7.10), including (*i*) *C.4.5* decision tree-based algorithm, (*ii*) *random forest*, and (*iii*) *RSM* approach. At first, it is clear that the basic classifiers, like *OneR* or *ZeroR*, are inefficient and useless in the process of mining *miRNA* data. Unexpectedly, results of *Naive Bayes* classifier were also weak. Furthermore, none of the tested decision-rule based classifiers was better than our new *LEMRG* algorithm which induces *cumulative* set of decision rules. On the other hand, very good results of classification of single-type classifiers could be observed for the decision tree-based classifiers. However, besides the C4.5

**Table 7.10** The comparison of classification accuracy of various machine learning models for *COLON* and *OVARY* concepts

| Classifier | Number of correctly classified cases | |
|---|---|---|
| | COLON | OVARY |
| Decision rule-based classifiers | | |
| *DecisionTable* | 0 | 0 |
| *JRip* | 1 | 0 |
| *OneR* | 0 | 0 |
| *PART* | 1 | 2 |
| *ZeroR* | 0 | 0 |
| Decision-tree based classifiers | | |
| *DecisionStump* | 0 | 3 |
| *J48 (C4.5)* | 1 | 3 |
| *RandomTree* | 0 | 3 |
| *REPTree* | 0 | 3 |
| Other type classifiers | | |
| *Classification via Regression* | 1 | 2 |
| *KStar* | 1 | 2 |
| *Naive Bayes* | 0 | 1 |
| Aggregated classifiers | | |
| *AdaBoostM1* (baseline classifier: *DecisionStump*) | 0 | 3 |
| *Bagging* (baseline classifier: *REPTree*) | 1 | 2 |
| *LEMRG* (baseline classifier: *MLEM2*) | 1 | 3 |
| *RandomForest* | 1 | 3 |
| *RSM* (baseline classifier: *REPTree*) | 1 | 3 |

algorithm, classification accuracy for each of those algorithms was still worse than the results of our decision *rule generations*. The increase of classification accuracy could be achieved mainly by using the aggregated classifiers. For instance, the *REPTree* classifier predicted correctly all cases of ovary cancer, but one case of colon cancer was misclassified. Whereas, after using *RSM* algorithm with *REPTree* as baseline model, all cases for both concepts were classified correctly, i.e., the results were comparable with the results obtained for the *LEMRG* decision rules. Thus, based on the obtained results, it was proven that aggregated classifiers outperform single-type classifiers, especially for data sets containing small number of cases.

## 7.4 Summary and Conclusions

At first, it should be noticed that the classifiers induced by our newly developed *LEMRG* algorithm – i.e., saved in the form of *cumulative* set of decision rules – can be clearly understandable by humans, being even nonexperts of the *AI* domain. Thus, the induced decision rules can be easily verified, applied, and adapted by the physicians or specialists of bioinformatics. Furthermore, our decision rules allow to classify correctly all cases of two types of cancers, by using expression levels of 11 *miRNAs*, including five levels of *miRNAs* for *colon* and six *miRNAs* for *ovary* cancer, respectively. For all *miRNAs* used in the *LEMRG* decision rules, strong connections to the *colon* (Table 7.11) and *ovary* (Table 7.12) cancers have been uncovered and proved by the other studies [55–57, 59–66].

For instance, in reference to the *colon* cancer, Pellatt et al. identified 16 *miRNAs* (including *miR-192*) useful in analysis of human *colorectal* carcinoma and normal *colonic mucosa* [55]. The authors used random forest analysis and verified findings with logistic modeling in a detached data set, obtained from population-based studies of *colorectal* cancer conducted in *Utah and the Kaiser Permanente Medical Care Program*. Thus, it was proved that *miR-192-5p* plays an important role in discriminating between carcinoma and normal mucosa for *colon* cancer. The other study showed that *miR-1* entails to several biological processes of tumor cell, in its differentiation, proliferation, and apoptosis [56]. In brief, downregulation of *miR-1* is one of the most frequent events in various cancers, where for the human *colon* cancer *miR-1* was downregulated in 84.6 % of the tumors. This decrease is significantly correlated with its target gene (*MET*) overexpression, especially in metastatic tumors. Overexpression of *metastasis-associated in colon cancer 1* (*MACC1*) and downregulation of *miR-1* are associated with the *MET* overexpression. Thus, consistent with the suppressive role of *miR-1* – expressed in vitro in *colon* cancer cells – reduced *MET* levels and impaired *MET*-induced invasive growth have been seen [59]. In the other study,

Table 7.11 *MiRNAs* selected by *LEMRG*, with the known connection to the *colon* cancer

| Id | Human miRNA | Cancer connection [15, 54] |
|---|---|---|
| EAM232 | Hsa-miR-192 | The *miR-192-5p* is important in discriminating between carcinoma and normal mucosa for *colon* cancer [55] |
| EAM238 | Hsa-miR-1 | The study on the human *colon* cancer showed that *miR-1* is downregulated in 84.6 % of the tumors [56] |
| EAM241 | Hsa-miR-203a | The *miR-203a* is upregulated in *CRC* (*colorectal cancer*) [57] |
| EAM298 | Hsa-miR-194 | The *miR-194* is significantly downregulated in *CRC* (*colorectal cancer*) tissues and shows the most obviously inhibited effects on *colorectal* cell proliferation [58] |
| EAM358 | Hsa-miR-323a | The *miR-323a-3p* is downregulated in *CRC* (*colorectal cancer*) [57] |

**Table 7.12** *MiRNA*s selected by *LEMRG*, with the known connection to the *ovary* cancer

| Id | Human miRNA | Cancer connection [15, 54] |
|---|---|---|
| EAM203 | Hsa-miR-135a | The ubiquitous loss of *miR-135a* expression is a crucial mechanism for overexpression of homeobox *A10* (*HOXA10*) in *EOC* cells, which is implicated in epithelial *ovarian* carcinogenesis [60] |
| EAM208 | Hsa-miR-141 | The serum *miR-141* is able to discriminate the *ovarian* cancer patients from healthy controls [61] |
| EAM225 | Hsa-miR-18a | The *hsa-miR-18a* is highly upregulated in serous *ovarian* carcinoma [62, 63] |
| EAM233 | Hsa-miR-196a | Expressed from *HOX* gene clusters and target *HOX* genes. Then, mutation of *HOX* genes could cause cancers [64] |
| EAM317 | Hsa-miR-155 | Several types of *B cell lymphomas* have 10- to 30-fold higher copy numbers of *miR-155* than normal circulating *B* cells [65] |
| EAM342 | Hsa-miR-135b | The *miR-135b-3p* is overexpressed in *EOC* (*epithelial ovarian cancer*) [66] |

Hou et al. explored the expression profile of *miRNAs* in the response of the human *colon* cancer cells (*HT29*s) to *5-FU* (*5-fluorouracil-induced*) treatment and nutrient starvation. Based on using *miRNA* microarray analysis, 4 downregulated *miRNAs* including *miR-323a-3p* and 27 upregulated *miRNAs* including *miR-203a* under these both conditions were identified [57]. These 31 *miRNAs* having the potential to target genes are engaged in the process of autophagy regulation for the human *colon* cancer cells. Thus, authors proved the potential of the selected *miRNAs* to modulate autophagy in the *5-fluorouracil-induced*-based chemotherapy in reference to the *CRC* (*colorectal cancer*). Another novel prognostic predictor in *colorectal* cancer seems to be the *miR-194* [58]. It was shown that the decrease of *miR-194* expression is strongly correlated with a poor prognosis in reference to patients with *CRC* and engaged in tumorigenesis by the regulation of the *MAP4K4/c-Jun/MDM2* signaling pathway.

For the *ovary* cancer disease, it was proved that *miR-135a* may be predictive of *EOC* (*epithelial ovarian cancer*) [60]. Furthermore, in [61] it was shown that *miR-141* can be used as biomarker for *ovarian* cancer, which is the fifth leading cause of female death globally due to its low survival rates. *Gao and Wu* extracted serum samples from 74 *epithelial ovarian* cancer patients, 19 borderline *ovarian* cancer, and 50 healthy samples. To detect the prognostic value of *miR-141*, the *Kaplan-Meier* curve and the *long-rank* test were performed. It was found that *miR-141* was significantly elevated in the *epithelial ovarian* cancer patients compared to healthy ones. In addition, the relative expression level of *miR-141* showed an escalating trend from early stages to advanced stages. Thus, patients with low *miR-141* level achieved significantly higher 2-year survival rate. Furthermore, *Gao and Wu* suggested that (*i*) serum *miR-141* is able to discriminate the *ovarian* cancer patients from healthy ones, as well as (*ii*) *miR-141* may be treated as a predictive biomarker for *ovarian* cancer prognosis. Additionally,

Yekta et al. found that *Hsa-miR-196a* is located inside of *homeobox* (*HOX*) clusters and is regarded to target *HOX* genes. *HOX* genes play essential roles during normal development in oncogenesis. Some of the genes could cause cancer directly when altered by mutations [64]. In the case of *miR-155*, it was proved [65] that several types of *B* cell *lymphomas* have from 10- to 30-fold higher copy numbers of *miR-155* than normal circulating *B* cells. In the other study, Wang et al. used *qRT-PCR* (*quantitative real-time polymerase chain reaction*) assay to profile the *ovarian* cancer samples. Based on the performed analysis, they identified the 10-*miRNA* signature, which allowed to differentiate human *ovarian* cancer tissues from the normal ones with 97 % sensitivity and 92 % specificity, respectively. This 10-*miRNAs* sequence included *miR-135b-3p*, which had a significantly higher expression level in *ovarian* cancer tissues than for the normal tissues [66].

In the further research, we are planning to verify components of our newly developed *LEMRG* (*Learning from Examples Module based on Rule Generations*) algorithm. At the beginning of our research, we have assumed that the *LEMRG* algorithm has to be general enough to allow using as components (so-called baseline models) a few different, well-known algorithms of decision rule induction. At present, *LEM2* (and *MLEM2*) algorithms were used to inducing subsequent decision *rule generations* for the selected *miRNA* data set [1]. Further, two other algorithms, including *GTS* (*General-To-Specific*) and *AQ*, will be implemented and investigated. However, it should be noticed that due to the different characteristics of four selected algorithms of inducing baseline models (i.e., traditional decision rules), it will be necessary to adapt some other methods of input data set preprocessing. For instance, only *MLEM2* algorithm provides handling imperfect data sets. Thus, in order to standardize the procedure of further induction of decision rule generations with using *MLEM2*, *LEM2*, *GTS*, and AQ as components, some selected methods of handling data sets which are *incomplete*, *inconsistent* and data described by the *continuous* attributes must be adapted. Thus, after implementation of all four algorithms, they will be compared with each other, mainly in regard to the *classification error rate*. In our opinion, application of a few algorithms of decision rule induction should allow to develop an *optimal* (or *quasi-optimal*) classifier, especially well fitted to the analyzed *miRNA* data sets. Secondly, so far the specific approach of dividing learning data set into two new subsets (*local* and *global*) was examined. In the further research, other (perhaps more efficient) learning data set modification procedures should be developed and checked. At the final stage, the comparison of *accuracy* and *stability* of decision *rule generations*, induced by using our newly developed *LEMRG* algorithm with other well-known established methods of data mining and knowledge discovery in medical databases, will be performed.

**Acknowledgments** The research work of Łukasz Piątek on this chapter has been supported by *ANDREA* (*ActiveNanocoatedDRy-electrode forEegApplication*) EU-funded *FP7-PEOPLE Marie Curie Industry-AcademiaPartnerships andPathways* (*IAPP*) project.

# References

1. Lu J, Getz G, Miska EA, Alvarez-Saavedra E, Lamb J, Peck D, Sweet-Cordero A, Ebet BL, Mak RH, Ferrando AA, Downing JR, Jacks T, Horvitz HR, Golub TR (2005) MicroRNA expression profiles classify human cancers. Nature 435:834–838
2. Grzymała-Busse JW, Piątek, Ł (n.d.) LEMRG – decision rule generation algorithm for mining microRNA expression data. The preliminary research. In: Proc. of the 1st international conference of digital medicine & medical 3D printing, 17-19.06.2016, Nanjing, China
3. Chee M, Yang R, Hubbell E, Berno A, Huang XC, Stern D, Winkler J, Lockhart DJ, Morris MS, Fodor PA (1996) Assessing genetic information with high-density dna arrays. Science 274:610–614
4. Golub TR, Slonim DK, Tamayo P, Huard C, Gaasenbeek M, Mesirov JP, Coller H, Loh ML, Downing JR, Caligiuri MA, Bloomfield CD, Lander ES (1999) Molecular classification of cancer: class discovery and class prediction by gene expression monitoring. Science 286:531–537
5. Bontempi G (2007) A blocking strategy to improve gene selection for classification of gene expression data. IEEE/ACM Trans Comput biol Bioinforma 4:293–300
6. Lazar C, Taminau J, Maganck S, Steenhoff D, Coletta A, Molter C, de Schaetzen V, Duque R, Bersini H, Nowé A (2012) A survey on filter techniques for feature selection in gene expression microarray analysis. IEEE/ACM Trans Comput Biol Bioinforma 9:1106–1119
7. Huang J, Fang H, Fan X (2010) Decision forest for classification gene expression data. Comput Biol Med 40:698–707
8. Stiglic G, Rodriguez JJ, Kokol P (2010) Finding optimal classifiers for small feature sets in genomic and proteomics. Pattern Recognit Bioinforma Adv Neural Control Elsevier Neurocomputing 73:2346–2352
9. Chen AH, Tsau Y-W, Lin C-H (2010) Novel methods to identify biologically relevant genes for leukemia and prostate cancer from gene expression profiles. BMC Genomics 11:274. doi:10.1186/1471-2164-11-274
10. Nanni L, Lumini A (2011) Wavelet selection for disease classification by DNA microarray data. Expert Syst Appl 38:990–995
11. Chen AH, Lin C-H (2011) A novel support vector sampling technique to improve classification accuracy and to identify key genes of leukemia and prostate cancer. Expert Syst Appl 38:3209–3219
12. Huerta EB, Duval B, Hao J (2010) A hybrid LDA and genetic algorithm for gene selection and classification of microarray data. Neurocomputing 73:2375–2383
13. Ghorai S, Mukherjee A, Sengupta S, Dutta PK (2011) Cancer classification from gene expression data by NPPC ensemble. IEEE/ACM Trans Comput Biol Bioinform 8:659–671
14. Ambros V (2004) The function of animal microRNAs. Nature 431:350–355
15. Sanger Institute. http://www.mirbase.org
16. Kim VN, Nam JW (2006) Genomics on MicroRNA. Trends Genet 22:165–173
17. Harfe BD (2005) MicroRNAs in vertebrate development. Curr Opin Genet Dev 15:410–415
18. Bartel DP (2004) MicroRNA: genomic, biogenesis, mechanism, and function. Cell 116:281–297
19. McManus MT (2003) MicroRNAs and cancer. Semin Cancer Biol 13:253–258
20. Brown D, Shingara J, Keiger K, Shelton J, Lew K, Cannon B, Banks S, Wowk S, Byrom M, Cheng A, Wang X, Labourier E (2005) Cancer-related miRNAs uncovered by the mirVana miRNA microarray platform. Ambion Technotes Newsl 12:8–11

21. Tran DH, Ho TB, Pham TH, Satou K (2011) MicroRNA expression profiles for classification and analysis of tumor samples, special section on knowledge discovery, data mining and creativity support systems. IEICE Trans Inf Syst 94:416–422
22. Ibrahim R, Yousri NA, Ismail MA, El-Makky NM MiRNA and gene expression based cancer classification using self-learning and co-training approaches. In: Proc. of the IEEE international conference on bioinformatics and biomedicine (BIBM 2013), 18-21.12.2013, Shanghai, China, p. 495–498
23. Lowery AJ, Miller N, Devaney A, McNeill RE, Davoren PA, Lemetre C, Benes V, Schmidt S, Blake J, Ball G, Kerin MJ (2009) MicroRNA signatures predict oestrogen receptor, progesterone receptor and MER2/neu receptor status in breast cancer. Breast Cancer Res 11:R27. doi:10.1186/bcr2257
24. Fang J, Grzymała-Busse JW (2006) Mining of microRNA expression data – a rough set approach. In: Proc. of the 1st international conference on rough sets and knowledge technology (RSKT 2006). Springer, Berlin, p. 758–765
25. Breiman L (1996) Bagging predictors. Mach Learn 24:123–140
26. Schapire RE (1990) The strength of weak learnability. Mach Learn 5:197–227
27. Freund Y (1995) An adaptive version of the boost by majority algorithm. Inf Comput 121:256–285
28. Freund Y, Schapire RE Experiments with a new boosting algorithm. Proc. of the 13th international conference on machine learning (ICML 1996), 3-6.07.1997, Bari, Italy, pp 148–156
29. Freund Y, Schapire RE (1999) A short introduction to boosting. J Jpn Soc Artif Intell 14:771–780
30. Ho TK (1998) The random subspace method for constructing decision forests. IEEE Trans Pattern Anal Mach Intell 29:832–844
31. Ho TK (1995) Random decision forests. In: Proceedings of the 3rd international conference on document analysis and recognition, 14–16.08.1995, Montreal, QC, Canada, pp 278–282
32. Wang CW New ensemble machine learning method for classification and prediction on gene expression data. In: Proc. of the 28th IEEE EMBS annual international conference, 30.08-3.09.2006, New York, USA, p. 3478–3481
33. Pawlak Z (1982) Rough sets. Int J Comput Inf Sci 11:341–356
34. Grzymała-Busse JW (1992) LERS – a system for learning from examples based on rough sets. In: Słowiński R (ed) Intelligent decision support. Handbook of applications and advanced of rough set theory. Kluwer Academic Publishers, Dordrecht, pp 3–18
35. Grzymała-Busse JW MLEM2 – Discretization during rule induction. In: Proc. of the international conference on intelligent information processing and WEB mining systems, IIPWM 2003, Springer-Verlag, 02-05.06.2003, Zakopane (Poland), p. 499–50
36. Paja W (2008) Budowa optymalnych modeli uczenia na podstawie wtórnych źródeł wiedzy, PhD Thesis. AGH University of Science and Technology, Kraków (in Polish)
37. Grzymała-Busse JW (2005) Rule induction. In: Maimon O, Rokach L (eds) Chapter (13) in data mining and knowledge discovery handbook. Springer, New York, pp 277–294
38. Grzymała-Busse JW (2004) C3.4 discretization of numerical attributes. In: Klösgen W, Żytkow J (eds) Handbook on data mining and knowledge discovery. Oxford University Press, Oxford, pp 218–225
39. Grzymała-Busse JW (2004) Rough set approach to incomplete data. In: Prof. of the international conference on artificial inteligence and soft computing (ICAISC 2004), Zakopane, Poland, Lecture Notes in Artificial Intelligence, 3070, Springer-Verlag, p. 50–55
40. Pawlak Z, Grzymała-Busse JW, Słowiński R, Ziarko W (1995) Rough sets. Commun ACM 38:89–95
41. Żytkow JM (2002) Types and forms of knowledge: rules. In: Klösgen W, Żytkow JM (eds) Handbook of data mining and knowledge discovery. Oxford Press, Oxford, pp 51–54
42. Quinlan JR (1996) Bagging, boosting, and C4.5. In: Proc. of the 13th national conference on artificial intelligence, AAAI Press/MIT Press, Cambridge, MA, USA, pp 725–730

43. Quinlan JR (1993) C4.5, programs for empirical learning. Morgan Kaufmann Publishers, San Francisco
44. Mitchell TM (1997) Machine learning. McGraw-Hill, Inc, New York
45. Stefanowski J (1998) Rough set based rule induction techniques for classification problems. In: Proceeding of 6th European congress on intelligent techniques and soft computing, 7-10.09.1998, Aachen (Germany), 1 pp 109–113
46. An A (2003.,No.4–5) Learning classification rules from data. Int J Comput Math Appl 45:737–748
47. Grzymała-Busse JW (2006) Rough set strategies to data with missing attribute values. In: Lin TY, Ohsuga S, Liau CJ, Hu X (eds) Foundations and novel approaches in data mining, studies in computational intelligence, vol 9. Springer-Verlag, Heidelberg, pp 197–212
48. University of California, Irvine. http://archieve.ics.uci.edu.datasets.html
49. Michalski RS, Mozetic I, Hong J, Lavrac N The multi-purpose incremental system AQ15 and its testing application to three medical domains. Proc. of the AAAI'86, Philadelphia, USA, 1986, p. 1041–1045
50. Grzymała-Busse JW (1997) A new version of the rule induction system LERS. Fundam Informaticae 31:27–39
51. Hippe ZS (1997) Uczenie maszynowe – obiecującą strategią przetwarzania informacji w biznesie? Informatyka 4:27–31; 5:29–33 (in Polish)
52. Weiss S, Kulikowski CA (1991) Computer systems that learn: classification and prediction methods from statistics, neural nets, machine learning, and expert systems, Chapter How to Estimate the True Performance of a Learning System. Morgan Kaufmann Publishers, San Mateo, pp 17–49
53. Sharma TC, Jain M (2013) WEKA approach for comparative study of classification algorithms. Int J Adv Res Comput Commun Eng 2(4):1925–1931
54. miRNA Map. http://mirnamap.mbc.nctu.edu.tw
55. Pellatt DF, Stevens JR, Wolff RK, Mullany LE, Herrick JS, Samowitz W, Slattery ML (2016) Expression profiles of miRNA subsets distinguish human colorectal carnicoma and normal colonic mucosa. Clin Transl Gastroenterol 7(3), e152. doi:10.1038/ctg.2016.11
56. Han C, Yu Z, Duan Z, Kan Q (2014) Role of microRNA-1 in human cancer and its therapeutic potentials. Biomed Res Int 2014., Article ID 428371:11
57. Hou N, Han J, Li J, Liu Y, Qin Y, Ni L, Song T, Huang C (2014) MicroRNA profiling in human colon cancer cells during 5-fluorouracil-induced autophagy. PLoS One 9(12), e114779, https://doi.org/10.1371/journal.pone.0114779
58. Wang B, Shen Z, Gao Z, Zhao G, Wang C, Yang Y, Zhang J, Yan Y, Shen C, Jiang K, Te Y, Wang S (2015) MiR-194, commonly repressed in colorectal cancer, suppresses tumor growth by regulating the MAP 4K4/c-Jun/MDM2 signaling pathway. Cell Cycle 14(7):1046–1058. doi:10.1080/15384101.2015.1007767
59. Migliore C, Martin V, Leoni VP, Restivo A, Atzori L, Petrelli A, Isella C, Zorcolo L, Sarotto I, Casula G, Comoglio PM, Columbano A, Giordano S (2012) MiR-1 downregulation cooperates with MACC1 in promoting MET overexpression in human colon cancer. Clin Cancer Res 18:737–747
60. Tang W, Jiang Y, Mu X, Xu L, Cheng W, Wang X (2014) MiR-135a functions as a tumor suppressor in epithelial ovarian cancer and regulates HOXA10 expression. Cell Signal:1420–1426. doi:10.1016/j.cellsig.2014.03.002
61. Gao YC, Wu J (2015) MicroRNA-200c and microRNA-141 as potential diagnostic and prognostic biomarkers for ovarian cancer. Tumour Biol 36:4843–4850
62. Miles GD, Seiler M, Rodriguez L, Rajagopal G, Bhanot G (2012) Identifying microRNA/mRNA dysregulations in ovarian cancer. BioMed Cent Res Notes. doi:10.1186/1756-0500-5-164
63. Wyman SK, Parkin RK, Mitchell PS, Fritz BR, O'Briant K, Godwin AK, Urban N, Drescher CW, Knudsen BS, Tewari M (2009) Repertoire of microRNAs in epithelial ovarian cancer as

determined by next generation sequencing of small RNA cDNA libraries. Public Libr Sci One. doi:10.1371/journal.pone.0005311
64. Yekta S, Shih IH, Bartel DP (2004) MicroRNA-directed cleavage of HOXB8 mRNA. Science 304:594–596
65. Eis PS, Tam W, Sun L, Chadburn A, Li Z, Gomez MF, Lund E, Dahlberg JE (2005) Accumulation of miR-155 and BIC RNA in human B cell lymphomas. Proc Natl Acad Sci USA 102:3627–3632
66. Wang LMJ, Ren AM, Wu HF, Tan RY, Tu RQ (2014) A ten-microRNA signature identified from a genome-wide microRNA expression profiling in human epithelial ovarian cancer. Public Libr Sci One. http://dx.doi.org/10.1371/journal.pone.0096472

# Chapter 8
# Privacy Challenges of Genomic Big Data

**Hong Shen and Jian Ma**

**Abstract** With the rapid advancement of high-throughput DNA sequencing technologies, genomics has become a big data discipline where large-scale genetic information of human individuals can be obtained efficiently with low cost. However, such massive amount of personal genomic data creates tremendous challenge for privacy, especially given the emergence of direct-to-consumer (DTC) industry that provides genetic testing services. Here we review the recent development in genomic big data and its implications on privacy. We also discuss the current dilemmas and future challenges of genomic privacy.

**Keywords** Personal genomic data • Genomic privacy • Computational methods

## 8.1 Genomics: One of the Most Demanding Big Data Disciplines

DNA sequencing is critical in understanding the basic structure and function of the human genome by determining the exact nucleotides and their order in a DNA molecule [18]. It also provides the foundational view of the genetic basis for human diseases including cancer [9, 30]. The automated and parallelized approaches of Sanger sequencing [25] directly led to the success of the Human Genome Project [18] and the sequencing projects of other important model organisms for biomedical research (e.g., the mouse genome [31]). The availability of these whole genomes has provided scientists with unprecedented opportunities to make novel discoveries for genome architecture and function, genome evolution, and molecular bases of variations in the human population and disease mechanisms [9].

---

H. Shen (✉)
Department of Engineering and Public Policy, Carnegie Mellon University, Pittsburgh, PA, USA
e-mail: hongs@andrew.cmu.edu

J. Ma (✉)
Computational Biology Department, School of Computer Science, Carnegie Mellon University, Pittsburgh, PA, USA
e-mail: jianma@cs.cmu.edu

In the past decade, the development of faster, cheaper, and higher-throughput sequencing technologies has dramatically expanded the reach of genomic studies. These "next-generation sequencing" (NGS) technologies (as opposed to the Sanger sequencing which is considered as first-generation) have been one of the most disruptive technological advances [23]. One demonstration of this exciting technology development is the cost reduction in sequencing. In 2001, the cost of sequencing a million base pairs was about \$5,000; but it only costs about \$0.05 in the mid-2015 (http://www.genome.gov/sequencingcosts/). In other words, it costs less than \$5,000 to sequence an entire human genome with $30\times$ coverage, and this cost continues to drop dramatically, making personalized genomics possible for the general public in the very near future. Illumina sequencing [3] has been the main player in numerous NGS applications, ranging from whole-genome sequencing and whole-exome sequencing, to RNA sequencing, etc. In addition, several other sequencing technologies have been developed in recent years and are sometimes referred as "third-generation sequencing," with Pacific Biosciences (PacBio) single-molecule real-time (SMRT) technology [17] and the Oxford Nanopore's nanopore sequencing [2] (which produces handheld portable sequencers) being the most promising. It is inevitable that such exciting advancement will continue in the next few years, with the availability of even higher-throughput, lower-cost, and higher-quality sequencing technologies. Genomics has become one of the most demanding big data disciplines as it has been estimated that we would reach one zettabase (one trillion gigabase) of sequence per year with 100 million to 2 billion human genomes sequenced by the year 2025 [29].

The development of new sequencing technologies has enabled dramatic advancement in our understanding of the human genome and the genetic mechanisms for human diseases. For example, as of March 2016, the NHGRI-EBI GWAS Catalog (https://www.ebi.ac.uk/gwas/) [33] contains over 2400 studies and greater than 16,000 unique SNP associations with various human diseases including Alzheimer's, autism, diabetes, obesity, schizophrenia, and many other complex disorders. The 1000 Genomes Project has sequenced over 2,500 individual genomes from various human populations to have a comprehensive resource of human genetic variations [8]. The Cancer Genome Atlas (TCGA) project will characterize the whole genome about 11,000 tumor samples from more than 30 cancer types to understand the genome-wide somatic mutation profiles [4]. However, despite such rapid advancement in genome sequencing and large-scale genomic studies of personal genomes and human disease genomes, our understanding of the genome and genetic variation associated with diseases remains limited.

In the human genome, only less than 2% of the genomes code for protein (~20,000 protein-coding genes in human). From the ENCODE project [5], we know that a significant proportion of the rest 98% of the so-called noncoding regions of the human genome have regulatory elements that control the expression of protein-coding genes in various cellular conditions. It is now acknowledged that many human diseases (including cancer) are caused by mutations in these regulatory sequences in the noncoding regions of the human genome together with the changes in protein-coding genes [21, 32]. However, even though there are existing technologies that can identify putative regulatory elements, the exact functional roles of these regulatory elements remain largely elusive, e.g., what genes a specific regulatory element regulates in what kind of conditions and what type of quantitative connections are there between gene expression and genetic variation in regulatory elements. Epistasis, which refers to the interaction among genetic variants, further complicates the problem [24]. Therefore, our overall understanding of the human genome function is still at the very early stage. The richness of genomic data not only enables us to answer new questions about basic human biology and human diseases; the limitation and uncertainty of our current understanding of the data itself also tremendously increases the potential for misuse and misinterpretation of such massive amount of data. For example, it creates new challenges for researchers and clinicians who have to determine what findings should be returned [22]. Moreover, all these challenges have been further elevated by the emergence of the direct-to-consumer (DTC) industry for genetic testing services, a business model enabled both by the dropping cost of genomic sequencing and the rapid development of network technologies.

## 8.2 Computational Methods to Reveal Patient Participation in Genomic Studies

In the genomic medicine community, it is widely accepted that sharing data is fundamentally important for the advancement of research [6]. Recently, the Global Alliance for Genomics and Health (GA4GH) was formed with the goal to fully realize the potential of genomic medicine to advance human health (https://genomicsandhealth.org). However, due to the fact that large amount of data is constantly being generated from more human individuals, it naturally creates privacy concerns for the individuals that have been involved in various projects and data cohorts. In the past few years, using computational techniques, a number of studies have shown that many types of genomic data can raise much more serious privacy concerns than we used to think.

In Homer et al. [14], the authors showed that it was possible to determine if an individual was involved in a mixture using high-density SNP genotypic profile and suggested that summary statistics such as allele frequency is not sufficient to mask identity in GWAS. The authors measured the difference between the distances of allele frequencies with respect to the mean allele frequencies of a reference population and with respect to the mean allele frequencies of the genomic mixture. A distance measure $D(Y_{i,j})$ was defined as:

$$D(Y_{i,j}) = |Y_{i,j} - Pop_j| - |Y_{i,j} - M_j|$$

where the first term measures the difference between the reference population allele frequency $Pop_j$ and the allele frequency of individual $Y_{i,j}$ for each SNP $j$, and the second term measures the difference between the allele frequency of the mixture $M_j$ at SNP $j$ and the allele frequency of $Y_{i,j}$ for each SNP $j$ [14]. To test, one-sample $t$-test was used:

$$T(Y_i) = (E(D(Y_i)) - \mu_0)/(SD(D(Y_i))/\sqrt{s})$$

where $\mu_0$ is the mean of $D$ ($Y_{i,j}$) over individuals $Y_k$ that are not in the mixture, $SD$ ($D$ ($Y_i$)) is the standard deviation of $D$ ($Y_{i,j}$) for all SNP $j$ and individual $Y_i$, and $s$ is the total number of SNPs [14].

The study in Im et al. [16] demonstrated a method that can infer an individual's participation in a GWAS study when regression coefficients from quantitative phenotypes are available. The authors defined the following statistic:

$$\hat{Y}_I = \frac{n}{M} \sum_{j=1}^{M} \hat{\beta}_j (X_{I,j} - \hat{X}_j)$$

where $n$ is the number of individuals, $M$ is the number of independent SNPs, $X_{I,j}$ is defined as the allelic dosage of individual $I$ at SNP $j$, $\hat{\beta}_j$ denotes the estimated coefficient from fitting the model $Y_i = \alpha_j + \beta_j X_{i,j} + e_i$, and $\hat{X}_j$ is the estimated mean of allelic dosage for SNP $j$ from the reference group [16]. The authors proposed a statistical approach to estimate individual contribution to the regression coefficient.

The GA4GH has the Beacon Project in order to simplify data sharing through a web service called beacon that provides only allele presence information. Users can query these beacons at different locations to learn about the available genomic data. However, in Shringarpure et al. [28], the authors presented a method based on likelihood ratio test to identify membership in the beacon that could reveal phenotypic information about the participants. The method is independent of allele frequencies. The study demonstrated that beacons actually do not protect individual privacy and are open to re-identification attacks.

In Schadt et al. [27], a statistical approach was proposed to predict known SNP variants based only on gene expression data and published expression quantitative trait loci (eQTLs), which can be used to accurately identify individuals in a large cohort. The authors wanted to determine if a given individual who has been genotyped belongs to a study cohort ($C$) where the individual's gene expression data have been generated. The expression traits are $\{e_1,e_2,...,e_N\}$, and the cis-eQTLs corresponding to SNPs are $\{s_1,s_2,...,s_N\}$; both are from an independent training set ($T$) for which genotype and gene expression data are available. For given gene expression information $e_i$ for individual $i$, the method approximates the empirical distribution for the three genotypes of $s_i$ using the normal density function $\Phi_T(e_i|g_i)$, where $g_i$ represents the vector of genotypes at $s_i$ for individuals in $T$. In the study cohort $C$, only the gene expression data are observed. The goal is to predict the genotypes of the corresponding SNPs. In other words, we want to calculate the probability of a given vector of genotypes $g_i$ for individuals in $C$ at SNP $s_i$, given the vector of expression values $e_i$ for the corresponding gene. The probability is presented as $P_C(g_i|e_i)$. The authors used the Bayes theorem:

$$P_C(g_i|e_i) \sim P_C(e_i|g_i) \times P(g_i)$$

where $P(g_i)$ is the vector of population frequencies for the genotype vector $g_i$. For example, if we assume that gene $j$ and SNP $j$ form a strong *cis*-eQTL and that the SNP $j$ has three genotypes, *AA*, *Aa*, and *aa*, then for individual $i$, for instance, we can calculate:

$$p\left(g_{j,i} = Aa|e_{j,i}\right) = p\left(e_{j,i}|\mu_{j,Aa,T}, \sigma_{k,Aa,T}\right) \times p\left(Aa_{j,T}\right)$$

where $p(Aa_{j,T})$ is the genotype frequency of SNP $j$ in $T$ for genotype *Aa* and $p(e_{j,i}|\mu_{j,Aa,T}, \sigma_{k,Aa,T})$ is the density for the normal distribution with mean and variance indicated for genotype *Aa*. The probability $p(g_{j,i}=Aa|e_{j,i})$ reflects the probability that a given individual $i$ has the *Aa* genotype for SNP $j$ when the gene expression for gene $j$ is known.

Gymrek et al. [12] reported that even the surnames can be re-identified from personal genomes by looking at a type of variants called short tandem repeats on chromosome Y together with genetic genealogy databases. A recent method was developed to specifically focus on linking different datasets (e.g., genotype-phenotype associations) in order to reveal sensitive information (i.e., linking attack) [13]. These studies have suggested that new analytical approaches have been constantly revising our understanding of genomic privacy and our ability is limited to completely protect privacy in the era of genomic big data.

## 8.3 Implications of Genomic Privacy

The availability of data about a given individual in different datasets (both genotype data and non-DNA-based data sources) can create new privacy concerns if the new knowledge of correlations among such datasets is utilized to attack and de-identify these data sources. Such new knowledge of correlations among different data sources and the new knowledge about the connections among genetic variants themselves will continue to grow as our understandings of the basic function of the human genome and disease mechanisms becomes more complete.

An emerging new challenge to genomic privacy comes from the direct-to-consumer (DTC) genetic testing services, companies such as 23andMe now not only offer personal genetic testing directly to consumers at an affordable price, but they also create online social networking community for sharing such information and connecting with friends and family members [19]. The sharing of personal genetic information in such a wide range of databases can put individuals at high risk for privacy attacks. In particular, consumers may not be aware of the potential harm on privacy after sharing their genetic information. Consumers typically do not have sufficient knowledge of the implication of data sharing and the harms on various social consequences, including discriminations during employment, vulnerability in health insurance, revealing privacy on health and history of their family, and creating opportunities for malicious advertising based on identity theft using genetic information. In May 2008, GINA (Genetic Information Nondiscrimination Act) was enacted by the US Congress to prohibit the use of genetic information in employment discrimination and denying coverage or charging higher premiums in health insurance based on genetic predisposition. However, this does not apply to life insurance, disability insurance, and long-term care insurance. In addition, even with GINA, studies have shown that the general public remains very nervous about the usage of genetic testing data [1].

A further privacy concern involves the issue of ownership: Who owns the samples and the genomic information generated from these samples? In an investigation of the privacy policies of six major DTC companies, scholars found that only two of them – 23andMe and Family Tree DNA – clearly mentioned the issue of ownership, while others remained silent on this gray area [15]. The consequences of this unsolved issue are profound as the company, if granted the ownership right over its consumers' genomic data, can sell or transfer such data to other companies or institutions [11].

## 8.4 Dilemma of Genomic Privacy

The rapid advancement of genomic big data, its wide social implications through the emerging DTC industry, and the limited knowledge so far for interpretation and understanding the consequences, have jointly created a set of dilemmas. The escalating privacy concerns around personal genomic information – arguably the most private data about oneself and his/her families – have, to a certain extent, conflicted with the interests of the scientific community, the DTC industry, and probably also with individual consumers. In Fig. 8.1, we summarize the different components related to the issue of genomic privacy.

First, although large-scale genomic database is indispensable for the advancement of genomic research, in recent years we have witnessed growing uneasiness among research participants toward the largely unknown future use of their genomic information [22]. Partly in response to such uneasiness, a group of experimental models have been proposed to strengthen the trust and reciprocal relations between researchers and participants. For example, the Personal Genome Project (PGP) takes a radical approach to ask its volunteers to waive any privacy claims over their genomic data. The "health information altruist" model, on the other hand, grants participants more control and autonomy in the process by treating them as "partners" rather than "subjects." The generalizability of these "citizen science" approaches, however, remains to be seen, as they usually have a high requirement both on the educational level and on the trust level of their participants [22].

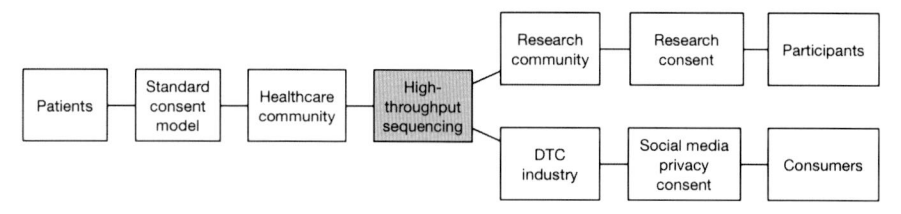

**Fig. 8.1** A summary of the different components related to the issue of genomic privacy

Second, the rise of the DTC industry and its social networking business model has also spurred public controversy, which triggered a series of regulations. For example, both the states of New York and California have tried to put a brake on the industry, sending cease-and-desist letters to companies in the field for not complying with state laws by requiring such tests to be offered with a physician's order. In the face of such conflicts, the future regulatory framework over DTC industry remains largely uncertain at the moment.

Finally, on the consumer end, some scholars and industry players have argued that consumers deserve the access to their own genomic and health information and have the right to share it. For example, in defending their positions against state regulations, major DTC companies have publicly argued that (1) people have a right to their genomic information, (2) the testing they are offering is not medical, and (3) patients should have "direct access to their health information without a physician intermediary" [20].

## 8.5 Current Limitations and Road Ahead

Currently no agreed-upon international regulations are available in this area, which will cause problems especially for developing countries. However, several international efforts are ongoing, including the Regulatory and Ethics Working Group (under GA4GH) which is focusing on ethical, legal, and social implications for global genomic data sharing, including the development of policies and standards, consent, privacy procedures, and best practices in data governance.

On the industrial level, although most DTC companies do provide privacy policies on their websites, a closer look at the consent process of major DTC companies reveals that most consent for the use of private information on these platforms is "presumed consent" [15]. This means that the providers presume that by reading the privacy policies and terms of service, consumers have understood and given their consent, unless they say otherwise, or "opt-out." Such "presumed consent" may cause varying degrees of privacy infringement. For example, people may not have enough knowledge of the process or consequences of such test, but they have very limited power to negotiate or question the company's policy if they want to use the service.

Another problem with the current industrial level approach lies in the largely undetermined, even unimaginable, future of genomic technology. As the privacy consent model used today in major DTC companies is based on and limited to the known technological application, future development in genomics might render such informed privacy consent futile [10].

Many different approaches have been proposed to better manage the current situation. Some advocate stronger state regulation. They argue that although personal genomic testing does not fit perfectly into the existing medical test regulatory model, genomic information still belongs to medical information, and future regulatory model should start from the standard medical model [20]. Others have taken an industrial approach. They worry that government regulation might be too overburdening and will create significant obstacle to industrial innovation while at the same time consumers should have rights to freely share their own genomic information [10]. Moreover, given the current unclear international regulatory framework and the rapid global expansion of DTC companies, an independent industrial approach might temporarily fill the regulatory void. Therefore, they urge providers in the DTC industry to form self-regulatory association and establish guidelines of best practices to protect consumers' privacy [11].

Still, others have questioned the notion of "privacy" itself. They argue that in the age of big data, our expectation of privacy has evolved quickly as the once clear line between private and public has become more and more blurred. Therefore, "education and legislation aimed less at protecting privacy and more at preventing discrimination will be key" [26]. Similarly, scholars have also advocated a trust-, instead of privacy-, centered framework. They argue that by enhancing transparency, control, and reciprocity in the research process, such as building a Bilateral Consent Framework that gives more power to participants on a global scale, the tension between data privacy and data sharing can be assuaged [7].

Although the future of genomic privacy remains largely unknown, it is clear that in the age of genomic big data, privacy protection development has lagged behind technological advancement. The policies and regulations remain largely passive rather than proactive. We believe that with the further diffusion of genomic technology, a collaborative framework between different stakeholders (policymakers, researchers, participants, industrial players, consumers) is necessary to advance the dialogue.

# References

1. Allain DC, Friedman S, Senter L (2012) Consumer awareness and attitudes about insurance discrimination post enactment of the Genetic Information Nondiscrimination Act. Fam Cancer 11:637–644
2. Ashton PM, Nair S, Dallman T, Rubino S, Rabsch W, Mwaigwisya S, Wain J, O'Grady J (2015) MinION nanopore sequencing identifies the position and structure of a bacterial antibiotic resistance island. Nat Biotechnol 33:296–300

3. Bentley DR, Balasubramanian S, Swerdlow HP, Smith GP, Milton J, Brown CG, Hall KP, Evers DJ, Barnes CL, Bignell HR et al (2008) Accurate whole human genome sequencing using reversible terminator chemistry. Nature 456:53–59
4. Cancer Genome Atlas Research N, Weinstein JN, Collisson EA, Mills GB, Shaw KR, Ozenberger BA, Ellrott K, Shmulevich I, Sander C, Stuart JM (2013) The cancer genome atlas pan-cancer analysis project. Nat Genet 45:1113–1120
5. Consortium EP (2012) An integrated encyclopedia of DNA elements in the human genome. Nature 489:57–74
6. Contreras JL (2015) NIH's genomic data sharing policy: timing and tradeoffs. Trends Genet 31:55–57
7. Erlich Y, Narayanan A (2014) Routes for breaching and protecting genetic privacy. Nat Rev Genet 15:409–421
8. Genomes Project C, Auton A, Brooks LD, Durbin RM, Garrison EP, Kang HM, Korbel JO, Marchini JL, Mc Carthy S, Mc Vean GA et al (2015) A global reference for human genetic variation. Nature 526:68–74
9. Green ED, Guyer MS (2011) Charting a course for genomic medicine from base pairs to bedside. Nature 470:204–213
10. Greenbaum D, Du J, Gerstein M (2008) Genomic anonymity: have we already lost it? Am J Bioeth 8:71–74
11. Gurwitz D, Bregman-Eschet Y (2009) Personal genomics services: whose genomes? Eur J Hum Genet 17:883–889
12. Gymrek M, McGuire AL, Golan D, Halperin E, Erlich Y (2013) Identifying personal genomes by surname inference. Science 339:321–324
13. Harmanci A, Gerstein M (2016) Quantification of private information leakage from phenotype-genotype data: linking attacks. Nat Methods 13:251–256
14. Homer N, Szelinger S, Redman M, Duggan D, Tembe W, Muehling J, Pearson JV, Stephan DA, Nelson SF, Craig DW (2008) Resolving individuals contributing trace amounts of DNA to highly complex mixtures using high-density SNP genotyping microarrays. PLoS Genet 4: e1000167
15. Huang H-Y, Bashir M. 2015 Direct-to-consumer genetic testing: contextual privacy predicament. In: Proceedings of the 78th ASIS&T Annual Meeting: information science with impact: research in and for the community, p. 50. American Society for Information Science
16. Im HK, Gamazon ER, Nicolae DL, Cox NJ (2012) On sharing quantitative trait GWAS results in an era of multiple-omics data and the limits of genomic privacy. Am J Hum Genet 90:591–598
17. Korlach J, Bjornson KP, Chaudhuri BP, Cicero RL, Flusberg BA, Gray JJ, Holden D, Saxena R, Wegener J, Turner SW (2010) Real-time DNA sequencing from single polymerase molecules. Methods Enzymol 472:431–455
18. Lander ES, Linton LM, Birren B, Nusbaum C, Zody MC, Baldwin J, Devon K, Dewar K, Doyle M, FitzHugh W et al (2001) Initial sequencing and analysis of the human genome. Nature 409:860–921
19. Lee SS, Crawley L (2009) Research 2.0: social networking and direct-to-consumer (DTC) genomics. Am J Bioeth 9:35–44
20. Magnus D, Cho MK, Cook-Deegan R (2009) Direct-to-consumer genetic tests: beyond medical regulation? Genome Med 1:17
21. Maurano MT, Humbert R, Rynes E, Thurman RE, Haugen E, Wang H, Reynolds AP, Sandstrom R, Qu H, Brody J et al (2012) Systematic localization of common disease-associated variation in regulatory DNA. Science 337:1190–1195
22. McEwen JE, Boyer JT, Sun KY (2013) Evolving approaches to the ethical management of genomic data. Trends Genet 29:375–382
23. Metzker ML (2010) Sequencing technologies – the next generation. Nat Rev Genet 11:31–46
24. Phillips PC (2008) Epistasis – the essential role of gene interactions in the structure and evolution of genetic systems. Nat Rev Genet 9:855–867

25. Sanger F, Nicklen S, Coulson AR (1977) DNA sequencing with chain-terminating inhibitors. Proc Natl Acad Sci U S A74:5463–5467
26. Schadt EE (2012) The changing privacy landscape in the era of big data. Mol Syst Biol 8:612
27. Schadt EE, Woo S, Hao K (2012) Bayesian method to predict individual SNP genotypes from gene expression data. Nat Genet 44:603–608
28. Shringarpure SS, Bustamante CD (2015) Privacy risks from genomic data-sharing beacons. Am J Hum Genet 97:631–646
29. Stephens ZD, Lee SY, Faghri F, Campbell RH, Zhai C, Efron MJ, Iyer R, Schatz MC, Sinha S, Robinson GE (2015) Big data: astronomical or genomical? PLoS Biol 13:e1002195
30. Vogelstein B, Papadopoulos N, Velculescu VE, Zhou S, Diaz LA Jr, Kinzler KW (2013) Cancer genome landscapes. Science 339:1546–1558
31. Waterston RH, Lindblad-Toh K, Birney E, Rogers J, Abril JF, Agarwal P, Agarwala R, Ainscough R, Alexandersson M, An P et al (2002) Initial sequencing and comparative analysis of the mouse genome. Nature 420:520–562
32. Weinhold N, Jacobsen A, Schultz N, Sander C, Lee W (2014) Genome-wide analysis of noncoding regulatory mutations in cancer. Nat Genet 46:1160–1165
33. Welter D, MacArthur J, Morales J, Burdett T, Hall P, Junkins H, Klemm A, Flicek P, Manolio T, Hindorff L et al (2014) The NHGRI GWAS catalog, a curated resource of SNP-trait associations. Nucleic Acids Res 42:D1001–D1006

# Chapter 9
# Systems Health: A Transition from Disease Management Toward Health Promotion

Li Shen, Benchen Ye, Huimin Sun, Yuxin Lin, Herman van Wietmarschen, and Bairong Shen

**Abstract** To date, most of the chronic diseases such as cancer, cardiovascular disease, and diabetes, are the leading cause of death. Current strategies toward disease treatment, e.g., risk prediction and target therapy, still have limitations for precision medicine due to the dynamic and complex nature of health. Interactions among genetics, lifestyle, and surrounding environments have nonnegligible effects on disease evolution. Thus a transition in health-care area is urgently needed to address the hysteresis of diagnosis and stabilize the increasing health-care costs. In this chapter, we explored new insights in the field of health promotion and introduced the integration of systems theories with health science and clinical practice. On the basis of systems biology and systems medicine, a novel concept called "systems health" was comprehensively advocated. Two types of bioinformatics models, i.e., causal loop diagram and quantitative model, were selected as examples for further illumination. Translational applications of these models in systems health were sequentially discussed. Moreover, we highlighted the bridging of ancient and modern views toward health and put forward a proposition for citizen science and citizen empowerment in health promotion.

**Keywords** Systems health • Systems biology • Complexity • Critical transitions

L. Shen • B. Ye • Y. Lin • B. Shen (✉)
Center for Systems Biology, Soochow University, No.1 Shizi Street, Suzhou, Jiangsu 215006, China
e-mail: bairong.shen@suda.edu.cn

H. Sun
Collaborative Innovation Center of Sustainable Forestry in Southern China of Jiangsu Province, Nanjing Forestry University, Nanjing 210037, China

H. van Wietmarschen (✉)
Louis Bolk Institute, 3972, LA, Driebergen, The Netherlands
e-mail: H.vanWietmarschen@Louisbolk.nl

## 9.1 Introduction

Over the last several decades, the pace of life is rapidly increasing. A modern lifestyle is often characterized by demanding jobs, families in which both parents work and care for children and other family members, and a busy social life. Attempts to sustain this kind of lifestyle are often accompanied by stress [1], late work hours, sleep problems [2], no time to cook, unhealthy diet [3], and too little exercise [4]. As the population ages, the amount of people suffering from lifestyle-related diseases such as cardiovascular disease, cancer, diabetes, COPD, and Alzheimer is soaring [5]. Unhealthy lifestyle combined with an aging population results in an increasing demand on health-care resources. Unfortunately, the current health-care system is failing to respond appropriately to meet this demand, resulting in high health-care cost in many countries, especially in developing countries. Scientists and doctors are working on several strategies in order to relieve the current pressure on the health-care system. However, health-care professionals themselves as well as medical students count among the highest numbers of burnout cases showing an inability to regulate their own stress levels adequately [6].

One strategy is further improving disease management. In 1996, Dr. Robert S. Epstein and Dr. Louis M. Sherwood clearly put forward the definition of disease management that disease management refers to the use of an explicit systematic population-based approach to identify persons at risk, intervene with specific programs of care, and measure clinical outcomes [7]. Since then, disease management has become an effective way for medical care. Doctors give patients advice on how to manage a chronic disease like diabetes and hypertension. Patients learn to take responsibility for understanding how to take care of themselves. They work together to avoid potential problems and exacerbation, or worsening, of their health problem. This was a quite effective approach, which increases patient satisfaction and improves their quality of life. However, there is no evidence that the primary goal of disease management, controlling health-care costs, is actually reached [8–10].

Another strategy developed in past decades is targeted therapy for complex diseases [11]. This brought chemical drug therapy into a new age. One main reason for this change is that single-target therapy can effectively relieve the side effect which usually appears in traditional broad-spectrum chemical drug therapy. Based on a large amount of OMICS data generated in recent years, as well as the rapid development of computer technology, bioinformatics, and cheminformatics, computer-aided drug design (CADD) has become an efficient tool for novel specific drug design for corresponding targets [12]. More and more novel single-target drugs have been designed and have been commercialized. However, for more than 80 % of the diseases, there is still no effective therapy currently.

Most of the medical care programs are based on linear thinking. The disease management and single-target therapy mentioned above are good examples for this. One of the reasons that these strategies are not optimal for treating chronic diseases is that these strategies lack a consideration of the holistic nature of human beings.

Most chronic diseases are complex lifestyle-related conditions resulting from a disturbance in a combination of biological, psychological, social, spiritual, and environmental factors [13]. An optimal strategy for managing chronic diseases should therefore consider biology, psychology, and social environment together. More holistic approaches, sometimes called systems medicine approaches, have increasingly been developed and applied in the area of medicine over the last 10 years [14]. This has provided traditional medicine with new insights and guides us entering a new era of medical care. In recent years, more and more evidence is showing that our current health-care strategies have reached a tripping point. We need a shift from disease management toward existing knowledge, tools, and technology for health promotion. The aim of this review is to explore new insights in the area of health promotion and to introduce the integration of systems thinking with health-care science and clinical practice.

## 9.2 Complexity and Health

Here we divide the concept of health network into two parts, i.e., inner body network and environment network, for clear discussion. Our body can be considered as a complex system consisting various organs and units connected into one another. There is a widespread existence of self-regulating and adaptability mechanism at all levels and all part of the body. The interaction and mutual effect between related items can maintain the stability of human body vital signs and maintain the life activities of the body. One key reason why human body can deal with various external factors effectively is that our body is a dynamic system. Here we take 100K Wellness Project as an example. In 2014, Dr. Leroy Hood brought forward a new medical project in order to further promote P4 medicine [14, 15] (Fig. 9.1). The idea of this project is to take 100,000 well patients to carry out a detailed data collection in genomic, proteomic, metabolic, epigenetic, and phenotypic levels through examining patients' blood, saliva, stool, and other physiological and psychological factors. Only when every index in this examination stays in a reasonable range can patients present healthy situations. This project enhances the importance of precision medicine, and at the same time, it reflects the significance of system dynamic balance in human health. But there are two major drawbacks of this approach. The first one is the lack of measurements at various time-points and therefore a lack of insight into the changes of the system over time and the relationships between various changes over time. The second drawback is the limitation to the consideration of the inner body system only. For health promotion, we cannot ignore our surrounding environment (Fig. 9.2), such as our families, communities, workspace, as well as personal lifestyle. All of them are linked together. For example, if one person's friends and colleagues have positive lifestyle and willing to share with you, your lifestyle will be naturally changed over time. One's lifestyle plays a key role in personal health. Besides our social environment effects, natural environment effects also should be taken into account. Relevant

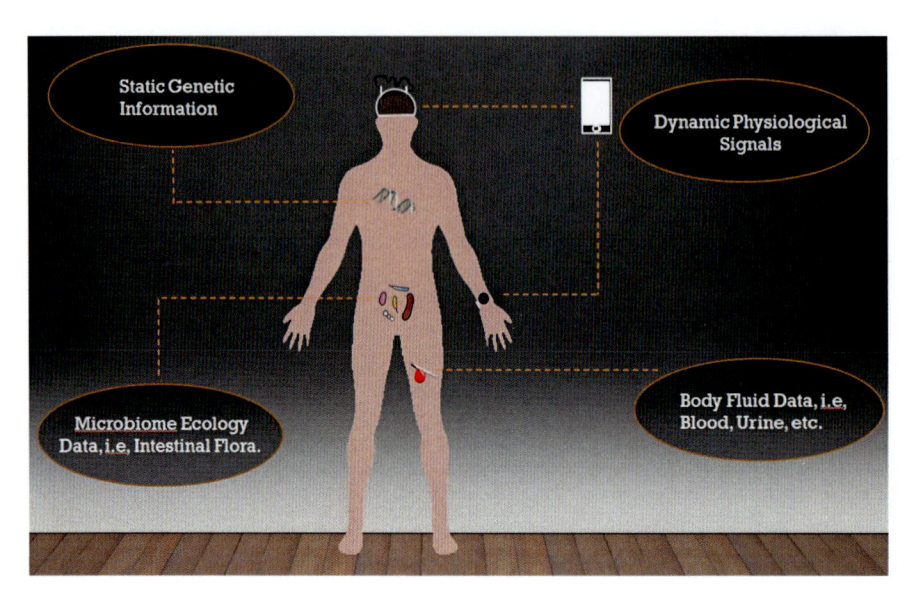

**Fig. 9.1** The paradigm of personal examination in 100K Wellness Project

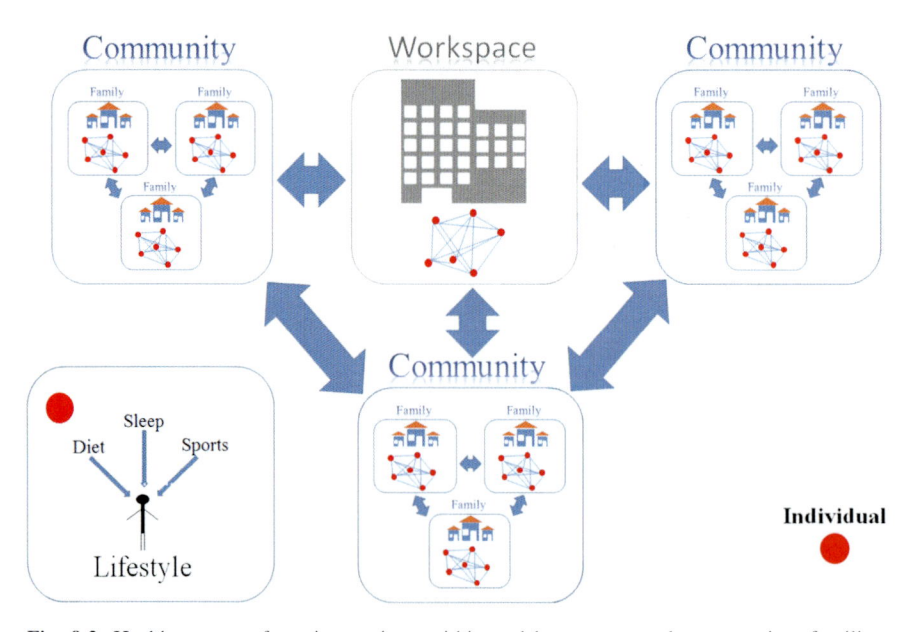

**Fig. 9.2** Health emerges from interactions within and between nested systems, i.e., families, communities, and workspace. The *red points* here represent individuals in these systems

researches have shown that many diseases, especially vector-borne diseases, are strongly associated with climate change [16]. Luis Fernando Chaves et al. found that cutaneous leishmaniasis, an emergent disease with increasing number of patients in the Americas, has a dynamic link with local climate cycles [17]. Researches in dengue fever, West Nile fever, and malaria also got the similar results and further support the linkage between disease and natural environment [18–21].

Complexity science offers methods to observe the dynamics of systems, as well as trajectories of changes within the dynamics [22]. It appears that lifestyle is developed over time into very stable habit patterns, which can be healthy patterns or patterns resulting in disease [23–25]. These habit patterns emerge from upbringing, role models, peer groups, social environment, and other factors [26]. This is one of the reasons that an unhealthy diet is so hard to change, despite large amounts of books about healthy diets [27]. Complexity science also shows that an unhealthy stable system might be changed into a healthier stable system through specific triggers. One well-known trigger is a life-changing event such as receiving a diagnosis of a life-threatening disease. Experiences from health coaches and patients indicate that during those times in life, people are suddenly realizing that they might have to change their lifestyle and are more open to make those changes. Complexity science offers theories and methods for studying such emerging habit patterns as well as critical transitions between various stable states of a system [28, 29].

## 9.3 From Systems Biology to Systems Health

In traditional biology research, biologists tend to specialize into fields covering limited parts of the human body even as small as individual cell systems, individual proteins, or individual metabolites. Even though our understanding of the mechanisms of biological processes and organ functions has increased, the importance of generating knowledge about the function of the human body as a whole is often overlooked by biologists and delegated to other fields of science such as philosophy, psychology, and sociology. Dr. Norbert Wiener, the founder of cybernetics, wanted to find out a new approach that could explain these processes in a holistic way. In 1948, he innovatively put forward an idea to illuminate complex bio-systems at a whole systems level [30]. Since then a new word called "biocybernetics" has been created, which is the predecessor of systems biology.

The aim of systems biology is to understand biological systems at a system level (Fig. 9.3). In other words, scientists investigate the components, structure, and dynamics of cellular and organismal function instead of the characteristics of isolated parts of a cell or organism [31]. Furthermore, systems biology develops tools, techniques, and models to predict the results brought about by stimulations of a living system combined with possible interferences from the outside environment. Based on systems theory, scientists and clinicians are able to develop holistic treatment strategies to complement simple molecular-targeted therapies. More and more systems theories are being applied to solve difficult medical issues and improve health care [32, 33]. This movement toward the implementation of systems

**Fig. 9.3** A schematic overview of the concept of systems biology. Generally it covers three research fields: biological science, information science, and systems science

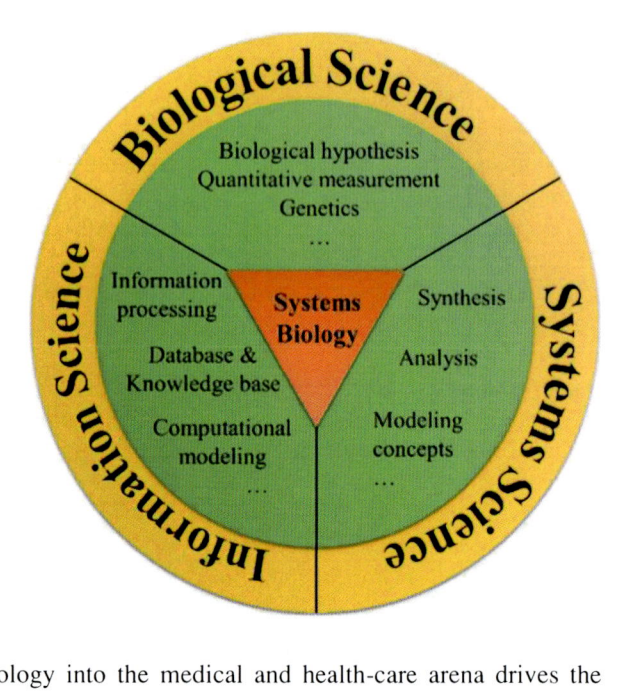

thinking and systems biology into the medical and health-care arena drives the development of what we call systems health, a biopsychosociospiritual approach toward health and healing [34]. An emerging field in systems medicine is the development of optimal healing environments, which are designed to optimize the environment in which patients are healing [35–37].

## 9.3.1 Network Biology

One of the science areas that is crucial for studying and understanding patterns of relationships is network biology [38]. Network biology arose from a merging of network theory, mathematics, and biology. In systems biology, network biology is commonly used to study changes in gene regulatory and protein interaction networks but also to explore relationships between social, psychological, environmental, and biological factors [39]. In other words, network biology reveals a phenomenon that relations among these items sometimes are more important than objects.

One example is the use of network biology to understand microRNA-mRNA interactions. MicroRNA (miRNA) is a small noncoding RNA containing 22–23 nucleotides. The first miRNA which was reported to play a role in gene regulation was found in 1993. Lee et al. noticed that a *Caenorhabditis elegans* gene, lin-4, which controls diverse postembryonic cell lineages, represses the expression of its target gene line-14 instead of encoding a protein [40]. Since then, an increasing number of researches have focused on human miRNA discovery, and considerable details are now known about their biogenesis [41]. Till now, more than 2,500 miRNAs have been annotated in humans [42–45], and in a recent study, miRNAs have been demonstrated to be effective biomarkers for complex diseases [46–48]. Based on this interaction, a human miRNA-mRNA regulation network has been built to further investigate the relationship among microRNAs and regulated genes. We can map those aberrant expressed genes or miRNAs onto this network to find out their relationships, helping us better understand the complex mechanisms [49].

### 9.3.2 Measuring Systems Health

Based on systems biology, systems thinking, as well as health theories, "systems health" was coined as a new frame of thinking about health as a dynamic complex system of relationships [34]. The concept of systems health challenges the manner in which health is currently assessed and monitored. Nowadays, measurement is mostly related to the absence of health or a disease state. For instance, cholesterol levels are measured in blood to detect hypertension, blood sugar levels are measured to indicate diabetes, and inflammatory markers are measured to detect inflammation or infections. However, these measurements do not tell much about the healthy state of the human system. Plenty of researches showed that health is not so much a state as an ability to adapt in the face of the challenges of life [50–52]. Measuring health therefore requires dynamic measurements to capture this ability to adapt this resilience. Furthermore, to capture a broad picture of someone's health, not only biological factors should be measured but psychological, social, and spiritual factors as well. This requires an ability to integrate data from various sources and methods to analyze this data as a whole [53, 54].

### 9.3.3 Systems Health Models

Mathematical and physical models are especially suitable to study the dynamics of systems, and there has been an increasing interest in such models in the field of systems biology over the last two decades [55]. Models are interesting for several reasons. First of all, systems health models can be used to increase understanding of how health emerges from relationships; it can help to raise health awareness. Furthermore, semiquantitative models can be used for clinical decision support, and quantitative models can be developed to study trigger points for behavior change. In the following section, we will mainly explore the first and third examples to further illuminate the value of systems health models.

### 9.3.3.1 Causal Loop Diagrams to Promote Health Awareness

Today, the majority of information about health and disease reaches people through social media, newspapers, magazines, television talking, websites, and social networking sites. However, there are many conflicting views on what is healthy and what is not. There are views on how to treat certain conditions and many sources of individual experiences with diseases and treatment options. This amount of information causes a lot of confusion for the majority of people because there is little integration of all the information into a coherent view on health. There is a great need for tools and methodologies to integrate the dynamics and the multitude of interacting factors related to health in a comprehensible way that can be used to generate health awareness [56–59].

Systems dynamics is a field of research which provides specific methods to describe causal relationships between factors into a model structure and provide options for simulating the dynamics involved in health and disease. An example of a causal loop diagram is shown in Fig. 9.4, a practical method to show causal relationships [60]. The systems health causal loop diagram shown in Fig. 9.4 was built by the Netherlands Organization for Applied Scientific Research (TNO) and discussed in a recent publication [34]. A large amount of expert knowledge was collected during focus group sessions form scientists in different research fields such as nutrition science, systems biology, computational modeling, and public health, to better understand the cross-domain relationships and to investigate the dynamical interactions between these domains in an integrated way. Another

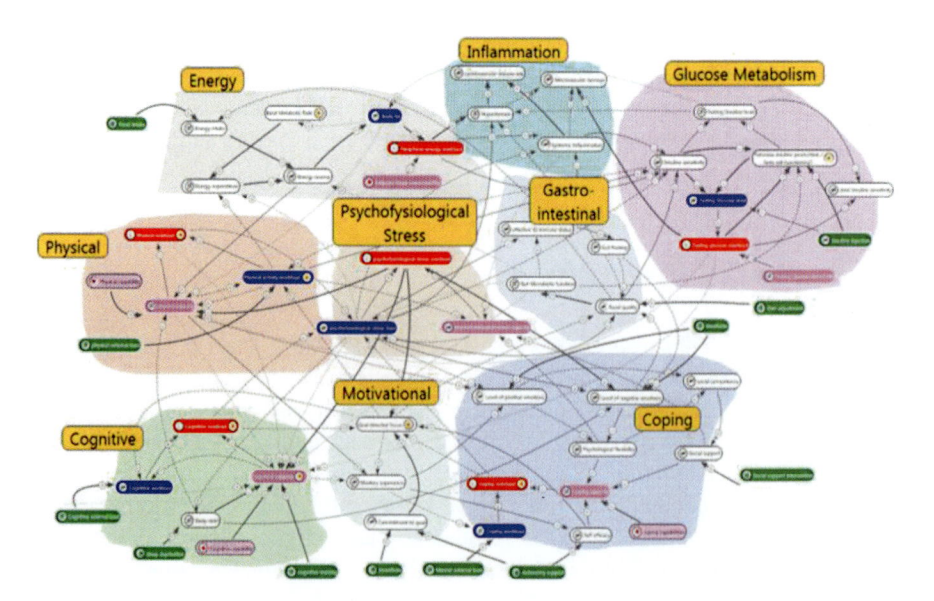

**Fig. 9.4** The causal loop diagram related to biopsychosocial health built by TNO to simulate intervention dynamics. Different domains are modularized in it. They are energy, inflammation, glucose metabolism, gastrointestinal, coping, motivational, cognitive, and physical [34]

example of a causal loop diagram application to health is from Jaimie McGlashan et al. [61]. In their research, they used network analysis combined with a causal loop diagram to find out the drivers of childhood obesity. Based on this, the researchers found keywords which were frequently discussed when it comes to children obesity, such as "Advertising/Sponsorship of Fast and Processed Food," "Level of Physical Activity," and "Participation in Sports." The feedback loops and reinforcing loops in the model help to connect different domains, from which the relationships between these phrases can be clarified and new insights about children obesity were generated. The examples above show that the causal loop diagram can be suitable tool for non-research people to better understand complex diseases in our life, enrich their knowledge in relevant fields, and further increase their health awareness (Fig. 9.4).

### 9.3.3.2 Quantitative Models to Discover Trigger Points for Behavior Change

Traditional mathematical modeling in biology research was usually used to inform and explain complex biological functions and processes and to understand the interactions between individuals through dynamic network based on gene expression and signaling pathways [62–64]. Generally these models are built based on theory and experimental observations. However, due to their small scales and the wile variations among individuals, plus the size of data storage which is rapidly increased, large-scale quantitative models are becoming increasingly necessary. Today, with the remarkable development of research technologies, a number of novel models have emerged, which allow us to assess the articulation of the changes of genes, proteins, and metabolites more comprehensively [65]. There is an obvious phenomenon that quantitative models are extending its application. In recent years, more and more quantitative models have been applied into clinical trials [66, 67], making a transition from fundamental research to medical application.

In order to illuminate quantitative models in more detail, here we select two investigations as examples. First is Marten Scheffer and his colleagues' work [68–70]. They developed a probabilistic model to compute the probability of symptom changes in major depression. The main formulas show as below:

$$A_i^t = \sum_{j=1}^{J} W_{ij} X_j^{t-1}$$

$$P(X_i^t = 1) = \frac{1}{1 + e^{(b_i - A_i^t)}}$$

Before developing this model, the authors made three assumptions: (1) symptoms $(X_i)$ can be on two conditions: active (1) or inactive (0), (2) symptom activation occurs $(t)$ over time, and (3) symptom $i$ and $j$ are connected with each other in the Virginia Adult Twin Study of Psychiatric and Substance Use Disorders (VATSPUD) data. Here, the first formula $A_i^t$ means total activation function. It represents that the total amount of activation symptom $i$ receives at time $t$ is the weighted ($W$) summation of all the neighboring symptoms X at time $t$-1. The second formula is the probability of the activation of symptom $i$ at time $t$. It depends on the difference between the threshold they estimated from VATSPUD data and $A_i^t$. The smaller the $|(b_i - A_i^t)|$ is, the more possible symptom $i$ will be active at time $t$. This is an intraindividual, symptoms-based model which develops over time. It gives a strong support to clinical doctors on major depression progression prediction and identification as it offers specified evaluation indexes that doctors are able to find out the trigger points and then distinguish different stages more easily.

The other good example for quantitative model application is Megan Sherwood et al.'s work [71]. They used meta-analytic and statistical approaches to examine the schizophrenia patients' response profile to clozapine, protracting a time course of symptom change. The formula they set is as below:

$$\text{Expected BPRS Item Score} = \beta_0 + \beta_1 {}^*W + \beta_2 {}^*W^2.$$

Here the $W$ represents week number. $\beta_0$, $\beta_1$, and $\beta_2$ are the coefficients they get from the regression analysis [72]. Through this model, they identified the range of weeks that distinguished different responses to clozapine. A supervising finding is that clozapine shows response in an early stage and the magnitude of it is somewhat larger than other antipsychotic drugs.

Meta-analysis approaches are pretty common in current clinical research [73–75]. But this is just an epitome of quantitative models. In other words, quantitative models have a great potential in medical application [76]. With huge amounts of medical data support, the demand for more accurate, more deep-level, and more complex models is growing [77]. With this method tend to mature, people are able to monitor their health systems in real time in the foreseeable future, which is an important object of systems health.

## 9.4 Bridging Ancient and Modern Views on Health

As Western medicine is struggling to move away from disease fighting toward health promotion [78], many traditional medicine systems are designed to promote health. In traditional Chinese medicine, a doctor who had to treat diseases was considered to be ordinary, whereas a doctor who treated the spirit or prevented diseases from occurring was considered an excellent expert. For ages TCM doctors were paid for the amount of healthy people in his care, which is a very different business model from the Western one, which pays for treatments [79]. Therefore there are many reasons to learn from traditional healing systems, not only for discovering new treatment options but also for discovering ways to maintain health and even discover government structures that promote health [80, 81].

One of the challenges for the integration of traditional healing systems and Western medicine is the poetic terminology that is often used in traditional healing systems. Concepts such as qi, prana, meridians, and chakras are often frowned upon by Western trained doctors and scientists because these concepts do not fit in their current thinking paradigm. Over the last several decades, attempts have been made to translate concepts and ideas from Chinese medicine and Ayurveda to Western scientific thinking using systems biology [82, 83]. Several studies have been conducted to discover subtypes of chronic diseases such as rheumatoid arthritis, diabetes, and metabolic syndrome, based on TCM diagnostic patterns [84–86]. Moreover, these subtypes were related to biological pathways and mechanisms well known in Western science, opening up opportunities to use these subtypes in clinical practice [87]. One of the big challenges in translating Chinese medicine diagnosis is the inconsistency that is often encountered among various TCM doctors [88, 89]. Therefore various research groups are working on standardizing symptom patterns with questionnaires [90–92]. This type of research shows that it is possible to build a bridge between Western science and medicine and traditional healing systems.

## 9.5 Transition Toward Citizen Science and Citizen Empowerment

Over the past years, there is a rapid development of tools and devices to measure your own health, as well as apps to store and analyze health data [93]. People are going to have much more data and information about their own health. Interestingly, a lot of this data is related to symptoms that are frequently experienced by individuals. Ecological momentary assessment is a new area of research focusing on capturing symptom data in a simple manner but within the context of daily life [94]. Novel data analysis methods are being developed to analyze this type of personal health data, allowing an individual to monitor the effects of self-chosen interventions on health [95]. More advanced data analysis methods have been developed to predict changes in the dynamics of symptom patterns for migraine and depression, allowing a timely prediction of a migraine attack [96] or a depression episode [69]. Symptom patterns commonly used in traditional healing systems can be very interesting for such self-monitoring approaches. The patterns can be used to distinguish between types of migraine, depression, or stress and can, for example, be used to direct people toward effective dietary interventions [97].

Health data is going to be owned more and more by the individual instead of health institutes. Currently, individuals start organizing themselves in health data cooperatives, independent organizations that are responsible for protecting the privacy of the participants and can mediate in contributing the data to scientific projects [98]. A famous example is PatientsLikeMe, a network with over 500,000 patients donating data toward hundreds of scientific projects [99]. Furthermore, it functions as a social network which people use to communicate personal experiences with drugs, treatments, and lifestyle interventions, helping others in their personal journeys toward health and well-being. Community platforms such as PatientsLikeMe stimulate empowerment of patients and empowerment of entire communities to participate in health promotion and scientific research [100]. It actually allows people to create health [101]. A systems approach toward health and disease is essential for such communities, as it integrates all the aspects that are relevant for the life of a patient and citizen. It will help with the integration of biological, psychological, social, and environmental health measurements and guide citizens toward relevant interventions to sustain a healthy life.

**Acknowledgments** This study was supported by the National Natural Science Foundation of China (NSFC) (grant nos. 31670851, 31470821, and 91530320) and National Key R&D programs of China (2016YFC1306605).

# References

1. McEwen BS (2012) Brain on stress: how the social environment gets under the skin. Proc Natl Acad Sci U S A 109(Suppl 2):17180–17185
2. Irish LA et al (2015) The role of sleep hygiene in promoting public health: a review of empirical evidence. Sleep Med Rev 22:23–36
3. Lustig RH (2006) The 'skinny' on childhood obesity: how our western environment starves kids' brains. Pediatr Ann 35(12):898–902. 905–7
4. Kohl HW 3rd et al (2012) The pandemic of physical inactivity: global action for public health. Lancet 380(9838):294–305
5. Yach D et al (2004) The global burden of chronic diseases: overcoming impediments to prevention and control. JAMA 291(21):2616–2622
6. Dyrbye LN et al (2010) Burnout and serious thoughts of dropping out of medical school: a multi-institutional study. Acad Med 85(1):94–102
7. Epstein RS, Sherwood LM (1996) From outcomes research to disease management: a guide for the perplexed. Ann Intern Med 124(9):832–837
8. Mattke S, Seid M, Ma S (2007) Evidence for the effect of disease management: is $1 billion a year a good investment? Am J Manag Care 13(12):670–676
9. McCall N, Cromwell J (2011) Results of the medicare health support disease-management pilot program. N Engl J Med 365(18):1704–1712
10. Dewan NA et al (2011) Economic evaluation of a disease management program for chronic obstructive pulmonary disease. COPD 8(3):153–159
11. Gerber DE (2008) Targeted therapies: a new generation of cancer treatments. Am Fam Physician 77(3):311–319
12. Seoane JA et al (2013) Biomedical data integration in computational drug design and bioinformatics. Curr Comput Aided Drug Des 9(1):108–117
13. Wahlqvist ML (2014) Ecosystem health disorders – changing perspectives in clinical medicine and nutrition. Asia Pac J Clin Nutr 23(1):1–15
14. Flores M et al (2013) P4 medicine: how systems medicine will transform the healthcare sector and society. Perinat Med 10(6):565–576

15. Hood L, Lovejoy JC, Price ND (2015) Integrating big data and actionable health coaching to optimize wellness. BMC Med 13:4
16. Grasso M et al (2012) The health effects of climate change: a survey of recent quantitative research. Int J Environ Res Publ Health 9(5):1523–1547
17. Chaves LF, Pascual M (2006) Climate cycles and forecasts of cutaneous leishmaniasis, a nonstationary vector-borne disease. PLoS Med 3(8):e295
18. Akter R et al (2017) Joint effects of climate variability and socioecological factors on dengue transmission: epidemiological evidence. Trop Med Int Health 22(6):656–669
19. Samy AM et al (2016) Climate change influences on the global potential distribution of the mosquito Culex quinquefasciatus, vector of West Nile virus and lymphatic filariasis. PLoS One 11(10):e0163863
20. Ferrao JL, Mendes JM, Painho M (2017) Modelling the influence of climate on malaria occurrence in Chimoio municipality. Mozambique Parasit Vectors 10(1):260
21. Ivanescu L et al (2016) Climate change is increasing the risk of the reemergence of malaria in Romania. Biomed Res Int 2016:8560519
22. Holland JH (2006) Studying complex adaptive systems. J Syst Sci Complex 19(1):1–8
23. (1983) Health promotion: alcohol and drug misuse prevention. Publ Health Rep. Suppl: p 116–32
24. Leung DY et al (2016) Hardcore smoking after comprehensive smoke-free legislation and health warnings on cigarette packets in Hong Kong. Public Health 132:50–56
25. Schane RE, Glantz SA, Ling PM (2009) Social smoking implications for public health, clinical practice, and intervention research. Am J Prev Med 37(2):124–131
26. McEwen BS, Getz L (2013) Lifetime experiences, the brain and personalized medicine: an integrative perspective. Metabolism 62(Suppl 1):S20–S26
27. Adam TC, Epel ES (2007) Stress, eating and the reward system. Physiol Behav 91(4):449–458
28. Jayasinghe S (2012) Complexity science to conceptualize health and disease: is it relevant to clinical medicine? Mayo Clin Proc 87(4):314–319
29. Scheffer M et al (2012) Anticipating critical transitions. Science 338(6105):344–348
30. Wiener N (1948) Cybernetics. Sci Am 179(5):14–18
31. Kitano H (2002) Systems biology: a brief overview. Science 295(5560):1662–1664
32. Liu ET, Kuznetsov VA, Miller LD (2006) In the pursuit of complexity: systems medicine in cancer biology. Cancer Cell 9(4):245–247
33. Haase T et al (2016) Systems medicine as an emerging tool for cardiovascular genetics. Front Cardiovasc Med 3:27
34. van Wietmarschen HA, Wortelboer HM, van der Greef J (2016) Grip on health: a complex systems approach to transform health care. J Eval Clin Pract
35. Jonas WB, Chez RA (2004) Toward optimal healing environments in health care. J Altern Complement Med 10(Suppl 1):S1–S6
36. Freedman MR, Stern JS (2004) The role of optimal healing environments in the management of childhood obesity. J Altern Complement Med 10(Suppl 1):S231–S244
37. Schweitzer M, Gilpin L, Frampton S (2004) Healing spaces: elements of environmental design that make an impact on health. J Altern Complement Med 10(Suppl 1):S71–S83
38. Barabasi AL, Gulbahce N, Loscalzo J (2011) Network medicine: a network-based approach to human disease. Nat Rev Genet 12(1):56–68
39. Lucas M, Laplaze L, Bennett MJ (2011) Plant systems biology: network matters. Plant Cell Environ 34(4):535–553
40. Lee RC, Feinbaum RL, Ambros V (1993) The C. elegans heterochronic gene lin-4 encodes small RNAs with antisense complementarity to lin-14. Cell 75(5):843–854
41. Lin S, Gregory RI (2015) MicroRNA biogenesis pathways in cancer. Nat Rev Cancer 15(6):321–333
42. Hsu SD et al (2011) miRTarBase: a database curates experimentally validated microRNA-target interactions. Nucleic Acids Res 39(Database issue):D163–D169

43. Xiao F et al (2009) miRecords: an integrated resource for microRNA-target interactions. Nucleic Acids Res 37(Database issue):D105–D110
44. Kozomara A, Griffiths-Jones S (2014) miRBase: annotating high confidence microRNAs using deep sequencing data. Nucleic Acids Res 42(Database issue):D68–D73
45. Rennie W et al (2014) STarMir: a web server for prediction of microRNA binding sites. Nucleic Acids Res 42(Web Server issue):W114–W118
46. Shen L et al (2016) Knowledge-guided bioinformatics model for identifying autism spectrum disorder diagnostic MicroRNA biomarkers. Sci Rep 6:39663
47. Hayes J, Peruzzi PP, Lawler S (2014) MicroRNAs in cancer: biomarkers, functions and therapy. Trends Mol Med 20(8):460–469
48. de Boer HC et al (2013) Aspirin treatment hampers the use of plasma microRNA-126 as a biomarker for the progression of vascular disease. Eur Heart J 34(44):3451–3457
49. Li Y, Zhang Z (2015) Computational biology in microRNA. Wiley Interdiscip Rev RNA 6(4):435–452
50. Huber M et al (2011) How should we define health? BMJ 343:d4163
51. Ryff CD, Singer BH, Dienberg Love G (2004) Positive health: connecting well-being with biology. Philos Trans R Soc Lond Ser B Biol Sci 359(1449):1383–1394
52. Ryff C et al (2012) Varieties of resilience in MIDUS. Soc Personal Psychol Compass 6(11):792–806
53. Adourian A et al (2008) Correlation network analysis for data integration and biomarker selection. Mol BioSyst 4(3):249–259
54. Derous D et al (2015) Network-based integration of molecular and physiological data elucidates regulatory mechanisms underlying adaptation to high-fat diet. Genes Nutr 10(4):470
55. Posada D (2008) jModelTest: phylogenetic model averaging. Mol Biol Evol 25(7):1253–1256
56. Lan TS et al (2014) An investigation of factors affecting elementary school students' BMI values based on the system dynamics modeling. Comput Math Methods Med 2014:575424
57. Pere D (2017) Building physician competency in lifestyle medicine: a model for health improvement. Am J Prev Med 52(2):260–261
58. Mechanick JI, Zhao S, Garvey WT (2016) The adipokine-cardiovascular-lifestyle network: translation to clinical practice. J Am Coll Cardiol 68(16):1785–1803
59. Homer JB, Hirsch GB (2006) System dynamics modeling for public health: background and opportunities. Am J Publ Health 96(3):452–458
60. Goh YM et al (2012) Dynamics of safety performance and culture: a group model building approach. Accid Anal Prev 48:118–125
61. McGlashan J et al (2016) Quantifying a systems map: network analysis of a childhood obesity causal loop diagram. PLoS One 11(10):e0165459
62. Umulis DM et al (2010) Organism-scale modeling of early drosophila patterning via bone morphogenetic proteins. Dev Cell 18(2):260–274
63. Jaeger J et al (2004) Dynamic control of positional information in the early drosophila embryo. Nature 430(6997):368–371
64. Fakhouri WD et al (2010) Deciphering a transcriptional regulatory code: modeling short-range repression in the drosophila embryo. Mol Syst Biol 6:341
65. Kurz FT et al (2017) Network dynamics: quantitative analysis of complex behavior in metabolism, organelles, and cells, from experiments to models and back. Wiley Interdiscip Rev Syst Biol Med 9(1):e1352
66. Garcia-Martin A, Fernandez-Golfin C, Zamorano-Gomez JL (2014) New quantitative model of aortic valve in PreTAVI patients. Rev Esp Cardiol (Engl Ed) 67(6):488
67. Hongwei W et al (2010) Nonspecific biochemical changes under different health statuses and a quantitative model based on biological markers to evaluate systemic function in humans. Clin Lab 56(5–6):223–225

68. Cramer AO et al (2016) Major depression as a complex dynamic system. PLoS One 11(12): e0167490
69. van de Leemput IA et al (2014) Critical slowing down as early warning for the onset and termination of depression. Proc Natl Acad Sci U S A 111(1):87–92
70. Wichers M et al (2014) Reply to Bos and de Jonge: between-subject data do provide first empirical support for critical slowing down in depression. Proc Natl Acad Sci U S A 111(10): E879
71. Sherwood M, Thornton AE, Honer WG (2012) A quantitative review of the profile and time course of symptom change in schizophrenia treated with clozapine. J Psychopharmacol 26 (9):1175–1184
72. Sherwood M, Thornton AE, Honer WG (2006) A meta-analysis of profile and time-course of symptom change in acute schizophrenia treated with atypical antipsychotics. Int J Neuropsychopharmacol 9(3):357–366
73. Ratheesh A et al (2017) A systematic review and meta-analysis of prospective transition from major depression to bipolar disorder. Acta Psychiatr Scand 135(4):273–284
74. Fusar-Poli P et al (2012) Predicting psychosis: meta-analysis of transition outcomes in individuals at high clinical risk. Arch Gen Psychiatry 69(3):220–229
75. Schumacher M, Rucker G, Schwarzer G (2014) Meta-analysis and the surgeon general's report on smoking and health. N Engl J Med 370(2):186–188
76. Swanson KR, True LD, Murray JD (2003) On the use of quantitative modeling to help understand prostate-specific antigen dynamics and other medical problems. Am J Clin Pathol 119(1):14–17
77. Issa NT, Byers SW, Dakshanamurthy S (2014) Big data: the next frontier for innovation in therapeutics and healthcare. Expert Rev Clin Pharmacol 7(3):293–298
78. Lindstrom B, Eriksson M (2005) Salutogenesis. J Epidemiol Community Health 59 (6):440–442
79. Zhang C et al (2016) Integration of Chinese medicine with western medicine could lead to future medicine: molecular module medicine. Chin J Integr Med 22(4):243–250
80. Roberti di Sarsina P, Alivia M, Guadagni P (2012) Traditional, complementary and alternative medical systems and their contribution to personalisation, prediction and prevention in medicine-person-centred medicine. EPMA J 3(1):15
81. Verhoef MJ et al (2005) Complementary and alternative medicine whole systems research: beyond identification of inadequacies of the RCT. Complement Ther Med 13(3):206–212
82. Scheid V (2014) Convergent lines of descent: symptoms, patterns, constellations, and the emergent interface of systems biology and Chinese medicine. East Asian Sci Technol Soc 8 (1):107–139
83. Mukherjee PK, Venkatesh P, Ponnusankar S (2010) Ethnopharmacology and integrative medicine – let the history tell the future. J Ayurveda Integr Med 1(2):100–109
84. van Wietmarschen HA et al (2012) Characterization of rheumatoid arthritis subtypes using symptom profiles, clinical chemistry and metabolomics measurements. PLoS One 7(9): e44331
85. van Wietmarschen H et al (2009) Systems biology guided by Chinese medicine reveals new markers for sub-typing rheumatoid arthritis patients. J Clin Rheumatol 15(7):330–337
86. Wei H et al (2012) Urine metabolomics combined with the personalized diagnosis guided by Chinese medicine reveals subtypes of pre-diabetes. Mol BioSyst 8(5):1482–1491
87. Schroen Y et al (2015) Bridging the seen and the unseen: a systems pharmacology view of herbal medicine. Science 350(6262):S66–S69
88. Coeytaux RR et al (2006) Variability in the diagnosis and point selection for persons with frequent headache by traditional Chinese medicine acupuncturists. J Altern Complement Med 12(9):863–872
89. Zhang GG et al (2005) Variability in the traditional Chinese medicine (TCM) diagnoses and herbal prescriptions provided by three TCM practitioners for 40 patients with rheumatoid arthritis. J Altern Complement Med 11(3):415–421

90. Tan S et al (2005) Traditional Chinese medicine based subgrouping of irritable bowel syndrome patients. Am J Chin Med 33(3):365–379
91. Gadau M et al (2016) TCM pattern questionnaire for lateral elbow pain: development of an instrument via a Delphi process. Evid Based Complement Altern Med 2016:7034759
92. Fu TC et al (2016) Validation of a new simple scale to measure symptoms in heart failure from traditional Chinese medicine view: a cross-sectional questionnaire study. BMC Complement Altern Med 16:342
93. Ernsting C et al (2017) Using smartphones and health apps to change and manage health behaviors: a population-based survey. J Med Internet Res 19(4):e101
94. Bos FM, Schoevers RA, Aan het Rot M (2015) Experience sampling and ecological momentary assessment studies in psychopharmacology: a systematic review. Eur Neuropsychopharmacol 25(11):1853–1864
95. Maes IH et al (2015) Measuring health-related quality of life by experiences: the experience sampling method. Value Health 18(1):44–51
96. Dahlem MA et al (2015) Understanding migraine using dynamic network biomarkers. Cephalalgia 35(7):627–630
97. Schroen Y et al (2014) East is east and west is west, and never the twain shall meet? Science 346(6216):S10–S12
98. Hafen E, Kossmann D, Brand A (2014) Health data cooperatives – citizen empowerment. Methods Inf Med 53(2):82–86
99. Frost J, Massagli M (2009) PatientsLikeMe the case for a data-centered patient community and how ALS patients use the community to inform treatment decisions and manage pulmonary health. Chron Respir Dis 6(4):225–229
100. Den Broeder L et al (2016) Citizen science for public health. Health Promot Int
101. Cloninger CR (2013) Person-centered health promotion in chronic disease. Int J Pers Cent Med 3(1):5–12